LIB/LEND/002
3/97
UNIVERSITY OF
WOLVERHAMPTON
LR/LEND/002
THE HC
Dudley Learning
Ca
Du
Wol
D0587640
ONE

WP 0813081 7

WITHDRAWN

THE HOUSING CRISIS

Edited by
PETER MALPASS

First published by Croom Helm Ltd, 1986
Reprinted 1987
Reprinted by Routledge 1990
11 New Fetter Lane, London EC4P 4EE

Printed and bound in Great Britain by
Biddles Ltd, Guildford and King's Lynn

British Library Cataloguing in Publication Data

The Housing crisis.
1. Housing - Great Britain
I. Malpass, Peter
363.5'0941

ISBN 0-415-05154-1

CONTENTS

Contents

PREFACE AND ACKNOWLEDGEMENTS

In the early weeks of 1985 final year housing degree students at Bristol Polytechnic embarked upon a project (a 'housing forum') to investigate the gathering housing crisis in Britain. A series of guest speakers were invited to talk on various aspects of the crisis and in view of the growing revival of interest in housing issues it seemed appropriate to make their papers available to a wider audience. Chapters 4, 5, 7 and 8 are based on these talks, and additional material has been specifically commissioned or brought in from elsewhere to provide a more comprehensive coverage.

After years of government complacency about housing it is now becoming clear that there are growing shortages of suitable accommodation, especially in certain areas of high demand, and huge problems of disrepair in the existing stock of dwellings in almost all areas. The expansion of home ownership, upon which so much emphasis has been placed, is now being shown to be much less than the panacea that it has often been assumed to be. Up and down the country local authorities are finding that each year they have less to spend on new building for those who need to rent, less to spend on maintaining existing council houses, and less to spend on providing grants for improvement and repair of private houses. The severity of the situation in traditionally deprived areas such as west central Scotland has led fourteen authorities in the Strathclyde region to undertake a joint campaign to demand more investment in public sector housing. And even in a city like Bristol, with a public image of much greater comfort and affluence, the city council has recently produced a campaigning report arguing that £141 million is required to bring its own housing stock up to a satisfactory standard, and a further £100 million at least is required for the private sector. Yet in 1985-86 the authority had only £7.7 million to spend on repairing and improving the council stock, 40% *less* than the previous year. Similar examples, and worse, could be found throughout Britain, from the largest to the smallest authorities.

An obvious question to arise is what can be done to alleviate the housing crisis? Before anything can be done it is necessary both to recognise the extent and intensity of the problem, and to understand what kind of problem it is. These are essential preludes to effective action on an appropriate scale. This book is therefore mainly concerned with spelling out the various dimensions of current housing problems and their impact on different social groups. It is also about establishing these problems as *housing* problems, requiring housing responses. The book is not conceived as a blue-print for future action, although some suggestions are made about the kinds of policies that are required to alleviate the current situation. Its main purpose is to contribute to the revival of informed debate about housing in Britain, and in this spirit the various chapters should be seen as representing the varying perspectives of their individual authors.

As editor I wish to thank all the contributors for agreeing so readily to participate in putting the book together and also my colleague Geoff Winn who helped to organise the original student project. David Byrne's chapter is a revised version of a paper presented to a conference on 'Housing and Social Change: Building a New Agenda', held at SAUS, University of Bristol in December 1984. Chapter 9, on housing policy and young people, is a slightly revised version of an article which appeared in *Youth and Policy*, no. 11, Winter 1984-85; I am grateful to the editor for permission to reproduce it here.

Thanks also to Shirley Weeks for typing and retyping drafts with great speed and efficiency.

Peter Malpass

LIST OF CONTRIBUTORS

Peter Malpass is Senior Lecturer in Social Policy at Bristol Polytechnic.

Philip Leather and Alan Murie are respectively Lecturer and Senior Lecturer at the School for Advanced Urban Studies, University of Bristol.

Ted Cantle is Under Secretary (Housing and Public Works) at the Association of Metropolitan Authorities.

Mike Gibson is Head of the Department of Town Planing, Polytechnic of the South Bank, London.

Valerie Karn is Professor of Environmental Health and Housing at the University of Salford.

John Doling and Bruce Stafford work at the Centre for Urban and Regional Studies, University of Birmingham.

David Byrne is Lecturer in Social Administration at the University of Durham.

Christine Oldman teaches part-time at Leeds Polytechnic and the University of Leeds.

Chris Holmes is Director of CHAR, the campaign for single homeless people.

Chapter One

FROM COMPLACENCY TO CRISIS

Peter Malpass

Is There a Housing Crisis?

Despite the undeniable long term improvements in housing conditions in Britain there have been increasingly frequent references to a new housing crisis in the mid 1980s. Among the first to predict a growing shortage was the then recently established House of Commons Select Committee on the Environment which, in its first report in 1980, predicted a deficit approaching half a million dwellings by the middle of the decade. (1) Later two academic economists, Fleming and Nellis, published a paper which looked at the projections of house-building in the Labour Government's Green Paper on Housing Policy (1977) compared with actual building rates achieved up to 1981, and they too foresaw a gathering problem of shortage by the mid 80s. (2) Meanwhile the Association of Metropolitan Authorities (AMA) has produced a series of reports warning not only of housing shortage but also of growing problems of disrepair and unfitness in the older housing stock. (3) Gibson and Langstaff also drew attention to the emerging crisis in urban renewal. (4) Other writers to identify the impending housing crisis have been Ball (5) and Lansley, (6) plus a whole series of individuals and organisations submitting evidence to the Duke of Edinburgh's enquiry into British housing. (7)

The worsening housing situation can be measured in various ways. On close examination the apparent surplus of dwellings over households soon melts away, once allowance is made for factors such as empty houses that are derelict or undergoing improvement, the large number of empty houses in the process of exchange and the growing popularity of second homes. In 1981 Shelter estimated that there was a true national shortage of 830,000 dwellings. (8) In consequence local authority waiting lists are lengthening, and recorded homelessness is increasing. (9) At the same time the production of new houses to let by local authorities has fallen from 75,500 in 1979 to just 32,806 in the whole of Great Britain in 1983. (10) This level of new building is the lowest for sixty years (apart from the war years). To make matters worse for people in need of rented housing, the sale of over 500,000 council houses since 1979 must have reduced the flow of dwellings available for letting. In the past slum clearance provided a useful route into council housing for many low income households, but in 1982-83 only 21,000 dwellings were demolished in Great Britain, compared with 88,700 in 1972-73. (11) This is despite the existence of over one million occupied dwellings which are acknowledged to be unfit for human habitation. However it might be argued that rehabilitation, funded by improvement and repair grants, represents a preferable alternative to clearance. Unfortunately there has been no guarantee that grant expenditure is channelled towards helping households in the worst conditions, and the continued cuts in public expenditure on housing seem certain to make grants even harder to obtain. Meanwhile the numbers of dwellings in need of substantial repair continue to grow. (12)

Public expenditure on housing has been cut by more than 50% since 1978-79. (13) The cuts might not be seen as a serious matter if it could be shown that private builders were increasing their output and providing suitable accommodation for households who would previously have looked to the local authorities for help. However, what has occurred during the first half of the 1980s is a major slump in the building industry <u>as a whole</u>. Private sector output has failed to increase in line with the fall in public sector production: local authority completions in Great Britain fell by 57.5% between 1979 and 1983, but private sector production also fell, although by less than 1%.

The reduction in housebuilding has been an important component of the growing crisis in the building industry, in which well over 400,000 workers are believed to be unemployed. This is a useful reminder that contemporary housing problems affect the producers as well as the consumers. The condition of the building industry is now so debilitated that it would be unable rapidly to respond to a substantial increase in demand, thus raising the probability of sudden price increase. Some large building firms have reacted to the situation by restructuring their output in order to concentrate on the most profitable forms of provision. Ironically a growth area in the private sector in the 1980s has been sheltered housing for owner occupiers, even though a growing proportion of the restricted public sector output is also sheltered housing. At the same time, while local authority provision for families has been heavily cut back, the volume builders in the private sector are also withdrawing from the first time buyer market, in order to concentrate on the more profitable 'executive' housing market. (14) A key factor in the difficulties at the cheaper end of the new housing market, and the second hand market too, has been the very high levels of mortgage interest rates throughout the first half of the 1980s. In sharp contrast to the mid 1970s when high inflation caused real interest rates to be negative, giving borrowers a great advantage over lenders, in the mid 1980s the position is reversed, with interest rates at record levels in both nominal terms and real terms. All this points to the conclusion that there is a housing crisis in Britain today, and that, in summary, it is a crisis of quantity, quality and access. In other words, there is not enough housing, a significant proportion of it is of very poor, often declining, standard, and there are problems of obtaining entry to both the rented and owner occupied sectors.

Nevertheless, against this view it can be argued that to talk of a housing crisis in the mid 1980s is grossly inflamatory and alarmist. It can easily be shown by comparison with the early years of the century, or even the middle, that very considerable improvements have been made. By all the conventional indicators housing conditions have indeed improved: until the late 1960s there was a crude surplus of households over dwellings, but since then the position has been reversed; in just twenty years, 1951-71 statutory overcrowding fell by almost two thirds; whereas in 1951 in England and Wales 9.7 million households were regarded as living in conditions that were shared, overcrowded, unfit, or lacking in one or more of the basic amenities, by 1976 the figure had fallen to 2.7 million, in a larger total population; (15) since 1945 over 8 million dwellings have been added to the stock in England and Wales, while some 1.5 million unfit dwellings have been demolished in slum clearance schemes. Going beyond the conventional, and arguably outdated, indicates it is clear that the spread of central heating and better insulation have made homes warmer, cleaner and more comfortable than in the past.

These sorts of long term historical comparisons have obvious appeal for government ministers anxious to contain or reduce public expenditure in times of economic recession. However there is also some support for this perspective within the academic literature on housing. In a book published in 1982 Donnison and Ungerson wrote that, "...for the great majority of ordinary people, housing conditions in the simpler physical sense have never been so good or so equal." (16) And later in the same work they say that, "Although Britain is now one of the poorest countries in western Europe, she has completed a bigger slum-clearance programme and a bigger subsidised house-improvement programme than any other country in the world, and her standards are among the best." (17)

Further arguments to support this position can be derived from the growth of home ownership, which successive governments since the early 1950s have seen as socially and economically beneficial. Home ownership has more than doubled as a proportion of the housing stock in the last forty years, and now stands at about 63% in England and Wales (although substantially lower in Scotland). This achievement marks considerable progress towards a long-standing objective, described in a Labour Government White Paper in 1965 as 'a long term social advance.' (18) Alongside the growth of home ownership the private rented sector has continued to decline, but in recent years important changes in local authority housing (specifically legislation on homelessness and rent rebates/housing benefit) have made that tenure more accessible to people who must rent. A case can therefore be made to suggest that Britain now has a fully modernised tenure structure based on home ownership for the majority and council housing as the main tenure for people who cannot afford to buy.

Closer scrutiny, however, suggests that there is little substance in these arguments. As a response to the assertion that there is a housing crisis in the mid 1980s they are unconvincing, for two main reasons. First, the resort to long term historical comparisons obscures the fact that conditions are now getting worse for many people. For them it is little comfort to know that life was even more difficult forty or eighty years ago. Second, the modernisation of the tenure system is itself a cause of the housing crisis. It is only now becoming clear that the spread of home ownership does not necessarily solve the housing problem. Historically the housing problem was associated with the private rented sector; home ownership and council housing were seen as relatively unproblematic. Now, however, the owner occupied sector is much more socially diverse than in the past, and contains large numbers of low income purchasers, often living in poor quality housing, or dwellings in need of substantial repair. At the same time, the penetration of more low income tenants into council housing has co-incided with the ageing of the stock and the emergence of serious structural and repair problems.

Nevertheless, it is still possible to accept that there are serious and growing problems for many people, without having to accept that these amount to a housing crisis. It is not necessary, in order to deny that there is a housing crisis, to believe that *everything* is satisfactory and that *no-one* is experiencing difficulties with their housing. It can be argued that there will always be people who, for reasons of personal inadequacy, disadvantage or misfortune are likely to suffer housing deprivation, and that their vulnerability is heightened in periods of economic recession. According to this view talk of a *housing* crisis is misconceived since the people affected are generally members of economically marginal groups - such as the elderly, unemployed, unskilled workers, single parents and ethnic minorities - whose housing problems are a reflection of circumstances with causes which lie elsewhere. As Donnison and Ungerson put it, "It is no longer the mass of working people who suffer the worst hardships; it is a wide variety of unorganised and less popular minorities... Most housing problems are really problems of unemployment, poverty and inequality." (19) This perspective represents a real challenge to the view of the housing crisis as a fundamental problem, rooted in the structure of housing provision itself. It deserves to be taken seriously, not least because, as will be demonstrated below, it is a perspective which has been shared by successive governments in Britain for almost two decades.

Redefining the Housing Problem

The housing problem, as it was understood throughout the twentieth century right up to the late 1960s, was essentially about quantity, quality and price. There was an overall housing shortage, with the total number of households exceeding the number of dwellings available. This overall shortage concealed more serious problems in areas of high demand. The early development of Britain as an urban industrial society bequeathed to the twentieth century a huge stock of dwellings which were old, small, densely packed together, often poorly built originally and badly maintained subsequently. Apart from problems of structural instability and dampness, lack of heating, hot water and modern sanitation made many of these dwellings unfit for human habitation. The poor quality of nineteenth century working class housing, particularly that occupied by lower paid workers, reflected the significant gap between the cost of decent accommodation and the price that many workers could afford to pay.

Housing policy as it developed after the First World War, and particularly in the period 1945-1968, was designed to respond to these three components of the problem. It was geared towards high levels of production, the elimination of slum housing and the provision of subsidies in order to lower the price paid by consumers for housing of adequate standard. The emphasis changed over time, (20) with a concentration on reducing the shortage in the first ten years after the Second World War, and only from the mid-1950s was there a return to slum clearance on a large scale. Nevertheless throughout the period there was a consensus on the need to build large numbers of dwellings each year. The annual targets were gradually bid up by politicians in both the Labour and Conservative Parties: in 1951 the Conservative Minister of Housing, Harold Macmillan, promised to build 300,000 houses per year, but by 1965 the Labour Government was aiming at 500,000 per year by 1970 (and at the 1966 general election the Conservative manifesto talked of 500,000 by 1968 (21)). However, the high output period came to an end during the late 1960s; housing policy swung away from high output at just about the same time that production reached record levels, when in 1967 and 1968 over 400,000 dwellings were built.

Gibson and Langstaff have shown that it was in 1967 that housing ministers began to prepare the way for a departure from established policy. (22) Ministers started referring to the approaching end of the housing shortage and the probability of an overall surplus of dwellings over households within a very few years. It was argued that over much of the country, apart from the big urban centres, the time was approaching when public investment in new housing could be scaled down. At the same time, however, it was envisaged that there would be an expansion of rehabilitation and improvement of older housing. This was to be based on voluntary action in newly designated general improvement areas, and elsewhere. The new policy was set out in a White Paper (23) in April 1968 and carried into law in the Housing Act 1969.

The output of new local dwellings fell dramatically, from 181,467 in 1968 to just 88,148 in 1973, the lowest level since 1946. In the private sector output peaked at 200,438 in 1968 and then declined to a low point of 140,865 by 1974. This great switch in policy, away from high output towards a focus on rehabilitation of existing dwellings, was presented to the public as a positive development, reflecting both the easing of the housing shortage and the increasing unpopularity of comprehensive redevelopment. In the new situation, it was argued, there was no longer a national housing shortage, although there were admittedly serious _local_ shortages, and there was no longer a need for large scale slum clearance. In future emphasis would be placed on rehabilitation and gradual renewal.

In other words, successive governments in the late 1960s and early 1970s (i.e. both Labour and Conservative) presented an optimistic interpretation of the housing situation. Two of the three historic components of the housing problem were apparently substantially reduced in scale and intensity. It was during the Conservative government of 1970-74 that attention turned to the third component, the question of the price of housing. Whereas in the past it has been generally accepted that subsidies in the public sector were necessary to produce decent dwellings at rents that working class tenants could afford, in the early 1970s the government took a very different line. Now, apparently, far from housing being too expensive, tenants as a whole, in both public and private sectors, were not paying enough. The Housing Finance Act, 1972, was designed to extend the 'fair rent' principle throughout the private rented sector, replacing remaining very low controlled rents, and to introduce fair rents to the public sector. The whole trust of the Act was to raise rents generally, and to reduce the level of subsidy for council housing. Previous governments had tried to raise rent levels, but never with such determination or by such large amounts, and always in the context of a policy of high output. By the early 1970s the reform of housing finance was at the forefront of housing policy.

Thus, by the early 1970s the view of the housing

problem that had been widely accepted for several decades was being replaced by a much more optimistic perspective, which emphasised that the national shortage affecting a wide spectrum of social groups was nearly over, that the need for slum clearance on a large scale was gone and that the housing finance issue had become one of households paying too little for their housing rather than too much. The housing problem, which had been at the forefront of domestic British politics throughout the century was in the process of being redefined and pushed into the background.

Questions of quantity and quality were now subordinate to the reform of housing finance. Although the Housing Finance Act was highly contentious and was to a large extent repealed by the Labour Government in 1975, in fact Labour ministers shared the general view that the housing problem was considerably reduced in scale. However, it was not just a question of scale; the housing problem was being redefined as different in character as well as smaller in scale. Reference has already been made to suggestions that a series of local shortages had replaced the national shortage, and in a speech in 1975 Anthony Crosland (Secretary of State for the Environment) acknowledged the persistence of apparently intractable problems, which he described as, "serious, indeed often desperate, <u>residual</u> housing problems - particularly for the worst off in our society. We have overcrowding. We have homelessness. We have families living in squalid conditions." (24) (emphasis added).

This theme was continued in the Green Paper on Housing Policy, published in 1977. In their foreword to the Green Paper for England and Wales the Secretaries of State wrote:

> "...it is a commentary on our times that housing is often discussed as though things were getting worse, when the facts do not support this view. We are better housed as a nation than ever before; and our standards of housing seem to compare well with those of similar and more prosperous countries. This should give pause to critics who start with an assumption that present arrangements have served and are serving badly.
>
> Nevertheless, the rising standards of the great majority contrast sharply and starkly with those of people still living in poor or unsuitable housing. There are people who still live

in slums, in houses lacking basic amenities, in overcrowded conditions, or who have to share against their will and in difficult circumstances. There are those of our fellow-citizens with special housing needs, such as the elderly, the disabled and the handicapped. And there are those who are homeless. In addition there is the problem of whole neighbourhoods in our inner cities which are weighed down with concentrations of bad housing." (25)

The Green Paper went on to present an interpretation of the housing situation and housing policy which developed the theme that on the whole things were satisfactory and improving, but that it was necessary to be more selective in future, focusing resources on groups and localities with particular needs. On the issues of housing supply and quality the message in the Green Paper was that the crude surplus of dwellings over households and the recent reductions in unfit housing heralded an era of reduced public sector building. (26) This notion was picked up by some observers, for instance, by Malcolm Clarke, who in a paper entitled "Too Much Housing?" (27) argued that, "Britain's housing stock is now sufficient to shelter all its inhabitants, and the resources absorbed by house-building could be limited to repairs and replacement of those no longer fit for their function."

However most informed comment on the Green Paper was critical of its optimistic perspective and its failure to produce proposals for thorough-going reform. Stewart Lansley, for example, wrote that, "For all the publicity gained during its 2½ year life and all the analysis presented in its 700 pages, the housing policy review is almost entirely a non-event in terms of its policy recommendations, and the housing situation in future years will suffer in consequence." (28) Other commentators, such as Michael Harloe (29) and Stephen Merrett (30) also identified the complacent attitude underlying the analysis and conclusions of the Green Paper. Michael Ball later suggested that it represented, "the high point of post-war complacency over housing provision." (31)

There were four main indicators of this heightened level of complacency. Two of these, referring to the diminished scale of the housing problem and its residual character, have already been mentioned. The other two were, first, an apparent belief that there was no need for fundamental reform of the housing system, and second, an uncritical commitment to the capacity of an enlarged owner occupied sector to meet individual housing needs. The general impression created by the Green Paper was that there was no longer a serious housing problem and that existing policies were serving the country quite satisfactorily, apart from some relatively minor aspects where room for improvement was identified (such as the distribution of capital spending by local authorities and council house subsidy arrangements). In fact the Green Paper proposals on capital allocations and public sector subsidies opened the way for the massive cuts later imposed by the Conservatives after 1979. In practice, therefore, these reforms have turned out to be very significant.

However, the important point here is that rejection of fundamental reform really meant rejection of any reform of owner occupation (and in this sense the Labour Green Paper of 1977 was very similar to the Conservative White Paper of 1971, which preceded the Housing Finance Act). The Green Paper was unremittingly positive about home ownership and the desirability of further growth of this sector. The text contained repeated references to the intrinsic merits of home ownership, especially its capacity to impart freedom, independence and mobility, as well as financial advantage and psychological fulfilment to individual owners. Merrett commented that, "On the rare occasions when the prose of the Green Paper becomes lyrical it is always with reference to home ownership." (32) The overwhelmingly supportive attitude to home ownership inevitably led to acceptance of the existing inequitable framework of tax reliefs and rejection of any real reform of owner occupation. So uncritical was the Green Paper's analysis of, and support for, this tenure that Della Nevitt was prompted to accuse the authors of displaying an "almost fanatical belief in owner occupation as a cure for almost all housing problems." (33)

It is indeed highly complacent to believe in the efficacy of further growth in owner occupation, given that the only way to achieve significant expansion is to draw in more and more low income households. Nevertheless, this is precisely the course that was adopted by the Conservative Government after May 1979. The centrepiece of housing policy became the right for council tenants to buy their houses at substantial discounts. This, it was claimed, was a 'fundamental' right. (34) Speaking before the introduction of the right to buy the new Secretary of State, Michael Heseltine, promised that, "Dreams are going to come true for many more people." (35)

A measure of how far housing policy had changed in little more than a decade was, on the one hand, the importance attached to the sale of council houses, a policy for changing the ownership of existing dwellings, adding nothing at all to the total stock. And on the other hand, far from promising the electorate so many hundred thousand dwellings each year, Michael Heseltine steadfastly refused to predict how many dwellings would be built. (36) By the early 1980s output targets had ceased to be a measure of ministerial virility and had disappeared from the political agenda. The Government has, however, continued to promote the expansion of home ownership as the cure-all for housing problems, even extending to 60% the discount available to council tenants of long standing. For those people not lucky enough to be council tenants in desirable properties there have been a series of other initiatives designed to expand home ownership amongst low income households. (37)

The point of all this is to show that over a period now approaching twenty years governments of both major parties have displayed an increasingly complacent attitude towards housing. This has involved substantially reduced emphasis on new building and redevelopment, the latter being only partially offset by rehabilitation, growing emphasis on measures designed to alter the tenure of existing dwellings, and a redefinition of the housing problem. By focusing on the plight of 'special needs' groups it has been possible almost to define away the housing problem, to down-grade housing from a major social problem to a symptom of other problems such as poverty and unemployment. In other words, according to the view of successive governments, the housing problem as such has been largely solved, and what remains is a series of residual problems mostly affecting people whose vulnerability arises outside the housing system.

Meanwhile, however, powerful forces were at work challenging and undermining this official definition of the situation. In particular demographic changes, the ageing of the housing stock and the restructuring of housing tenure have contributed, over a run of years, to the development of the current housing crisis.

Demographic Change

Although the total population of Great Britain has been growing only slowly since the late 1970s (following four years of overall decline), the projected rate of growth in the period up to the mid 1990s is greater than at any time since 1971-72. (38) However it is not so much the total population but changes in its age and household structure which are important for housing. In terms of housing demand the key factor is how people organise themselves into separate households. Demand for housing can increase even within a static or falling total population if new households are forming faster than old ones dissolve. In Britain the number of one and two person households has increased steadily over a long period since the Second World War, from 42% of all households in 1961 to 54% in 1981. (39) This is in part a reflection of the preference for separate household status amongst both younger and older people. These are also age groups that are experiencing continued growth in numbers in the 1980s.

The peaking of the birth rate in the mid 1960s means that the numbers of people aged 20-24 will reach a peak in 1986, and remain high for the rest of the decade. The age group 25-29 will peak in the early 1990s. (40) The importance of this is that the mid to late 1980s is certain to be a period when maximum numbers of young people will be forming new households and hoping to secure separate homes of their own. Despite the low level of population growth, therefore, there is currently a high level of household formation, with consequent pressure on the housing system. Ermisch has pointed out that the actual numbers of households seeking separate accommodation depends on a number of factors, such as the level of real incomes and rents and house prices, but he estimates that purely demographic factors will add about 80,000 households each year throughout the 1980s. (41) Other factors might double the rate of increase in new households. (42) Ermisch estimates that in Great Britain by 1993 there will be 925,000 more households than in 1981 as a result of age distribution changes. (43) Official statistics estimate that for England and Wales there will be an increase of 700,000 households between 1986 and 1991. (44)

If the large numbers of young people embarking on their housing careers is the main factor underlying an increased rate of household formation and greater demand for housing, at the other end of the age range a rather different set of problems emerges. The total number of people over retirement age increased by over 40% between 1951 and 1981, (45) but is not expected to increase for the remainder of the century. However, within the population over retirement age there will be a significant shift towards fewer 'young elderly' and more 'old elderly', i.e. there will be almost 9% more people aged over 75 in 1988 than in 1983. (46) Given the long term trend towards more elderly people living alone in separate households and the current preference in social range left policy for elderly people to remain in the community rather than move into institutional care as they become more frail, there is a growing problem for many of them in securing and maintaining appropriate accommodation.

Amongst the youngest age group of adults the problem is essentially one of finding accommodation for newly formed households. On the other hand, amongst the elderly it is more a question of suitable homes for existing households. The importance of this can be better appreciated by considering the way in which demographic changes interact with the ageing housing stock and the restructuring of the tenure system.

The Ageing Housing Stock

Britain has a very large stock of old houses; well over 6 million, almost 30% of the total stock, were built before 1919 (in Wales the proportion is over 40%). (47) There is some variation from region to region, but in most parts around half of these dwellings were built in the high output period of the 1890s and early 1900s before the great slump in house building in the eight or ten years before the First World War. It has been characteristic of house building over the last hundred years that production has been subject to marked fluctuations as changing market conditions have produced a cycle of boom and slump. The importance of this historical tendency is that building booms in one period produce long term consequences, rather like the way a bulge in the birth rate at one point in time has social and economic implications over a long period.

In the mid 1980s the great majority of dwellings built between 1871 and 1914 are still standing, all of them over seventy years old and most of them over eighty years old. The houses produced in large numbers around the turn of the century are all getting old together, contributing to a need for large amounts of investment in major repair and modernisation work. At present replacement rates a new house built in 1985 will not be due for replacement for over 900 years, which means that existing older houses also have a very long life in front of them. However, houses do not last for ever, even if they are well maintained throughout their lives, which has not been the experience of most older dwellings in Britain.

The problem of ageing housing is accentuated in the mid to late 1980s by the high level of construction in the inter-war years. Over a million council houses and flats were built between the wars. A fifth of these are now over sixty years old, and more than three quarters are over fifty years old. For a long time council housing was predominantly modern, lacking any houses that could be described as old. The situation has now changed and the repair and modernisation of inter-war council houses has become a major issue for many local authorities. Private builders enjoyed boom conditions in the 1930s, with output exceeding 250,000 each year from 1934 to 1938. These houses represent the next great addition to the backlog of major repairs. Built in a fiercely competitive climate which, especially at the cheaper end of the market, encouraged cost cutting economies, many 1930 houses are already in need or rewiring, reroofing and generally refurbishing.

To sum up this section, the point to be established is that Britain faces a serious problem of maintaining and improving an ageing housing stock, in which 30% is over seventy years old, and to this must be added the increasingly urgent needs amongst the four million dwellings built between the wars. Who is responsible and where are the resources going to come from?

Tenure Restructuring

A development of the greatest significance for an understanding of housing in Britain in the nineteenth century is the long term restructuring of the tenure system. (48) Before the First World War about 90% of all households were tenants of private landlords, but in the mid 1980s this form of tenure accounts for probably less than 10%. Some 61% of households in Great Britain are now owner occupiers, and the remainder are tenants of local authorities and housing associations.

This restructuring process has had important beneficial effects. In particular the growth of council housing has helped to break the historic link between poverty and poor housing, and the emergence of owner occupation has produced for many owners greater security, greater freedom, better conditions, lower costs in the long term and accumulation of a capital asset of far larger value than could ever be secured by saving. However, there are certain disadvantageous effects of tenure change. First, the decline of private renting has greatly reduced the availability of a form of housing with cheap and ready access. Council housing and owner occupation are tenures which present much greater barriers to entry. In the case of council housing these barriers are bureaucratic and require applicants to meet certain criteria and to attain priority within the council's allocation system. Typically this means queuing for months or years in order to obtain a council house. Owner occupation erects different sorts of access barriers, based on ability to pay. Again it is typically necessary for would-be home owners to wait some time while they accumulate a deposit before obtaining access to their preferred form of housing. The tenure restructuring process has, therefore, made access to decent housing more difficult in certain respects, especially for low income people and those who do not conform to the kinds of households (two parents and dependent children) who continue to be given priority by local authority housing departments. In this sense the growing problem of homelessness amongst single people is clearly due to changes in the system of housing provision and not a product of individual behaviour.

Second, home ownership has proved to be highly popular with consumers, producers and governments, but its continued growth has come to depend on absorption of more older, cheaper houses in inner city areas, and more purchasers on low incomes. This process has changed owner occupation into a much more diverse and socially mixed form of tenure, about which it is no longer safe to generalise on the desirability and financial advantages of ownership. In the past the dominant image of home ownership was of modern, suburban houses, occupied by relatively affluent families. Over time the growth and maturation of home ownership has drawn in a much wider range of dwellings, not all of which are selfevidently beneficial and appreciating capital assets, and a wider range of incomes and ages. Although home ownership can produce considerable financial gains for individual owners it also individualises costs: owners are solely responsible for repair and maintenance. In this sense low income and mortgaged house purchase, especially in run down inner city areas, do not fit comfortably together.

The point here is that tenure change has effectively removed the option of private renting on a long term basis for new households. The potential benefits of home ownership on the one hand and cutbacks in the provision of council housing on the other have both helped to persuade more and more low income households to opt for house purchase. Having done so these households face the prospect of mortgage costs that are high in relation to wages, leaving little or no capacity to pay for essential repair and maintenance work.

Another aspect of tenure change is that the rump of the traditional private rented sector now contains a high proportion of tenants who are elderly. Many of these are tenants of long standing whose living conditions constitute some of the worst housing to be found anywhere. However, in addition, the maturation of home ownership means that here too there are proportionately more elderly people than in the past. In its early years of expansion home ownership contained relatively few elderly people, but of course over time the younger people who bought houses in the 1930s boom have moved into retirement, and now an increasing proportion of post-war purchasers are also reaching this stage in their life cycle. Thus owner occupation has acquired a more balanced age structure, reflecting the proportion of elderly people in the population as a whole.

Elderly home owners do not usually face the heavy mortgage costs associated with younger low income purchasers, but they do face problems of repair and maintenance, to be financed out of retirement income. Obviously not all elderly home owners are poor but substantial numbers are dependent on state pensions and supplementary benefit or small occupational pensions. The difficulty they face is that their wealth is tied up in the house, and is therefore not easily accessible, but without expenditure on repair and maintenance the value of that asset may not be sustained. It is not just a question of available resources, however. No doubt the worry and disturbance associated with major repair or improvement work are, for many old people, sufficient deterrents in themselves. It may be that elderly people prefer the familiar surroundings of their established family home and do not notice any loss of amenity. But it may also be the case that the peculiarities of the British tax system, which give advantages to investment in housing, lead elderly people to remain as owners of unsuitably large dwellings rather than trade down to something small and modern, or even become tenants for their later years.

The Housing Crisis and Economic Policy

Preceding sections have shown that there are important long term developments affecting population structure, the ageing of the housing stock and the pattern of housing tenure, which individually influence the level of demand, the quality of housing available and the ease of access to decent accommodation at a reasonable price to the consumer. In combination these trends have had a much greater impact which has helped to produce the current housing crisis. For example, not only are there more elderly people than ever before, but also more of these elderly people are owner occupiers and more than ever before these elderly owners find themselves in ageing houses that they cannot afford to maintain or modernise. Similarly, the decline of private renting and cutbacks in the availability of council housing have pushed growing numbers of new or recently formed households into owner occupation, whether they can afford it or not. Policies designed to reduce the scale of comprehensive redevelopment and to promote home ownership inevitably mean that low income purchasers become the owners of, and responsible for, properties that are most in need of renovation.

A peculiar feature of the current housing crisis is its predictability, and yet at the same time the persistence of policies which have both denied its approach and made its impact more severe. It is almost as if official complacency has been inversely related to the condition of the economy - the worse the economy has become since the devaluation crisis of 1967, the greater the complacency about housing. Housing policy has been tailored to the growing difficulties of the economy, and the adoption of a complacent analysis of the housing problem has been a useful way of adjusting public expectations in line with what successive governments have deemed to be economically attainable. This can be seen as a kind of coping strategy (to borrow a social work term), a way of defining a situation that avoids the unpleasant consequences of facing up to reality. In this case facing up to reality over the last fifteen years or so would have meant sustaining a higher level of investment in new building and renovation, and recognising that continued expansion of home ownership to an ever increasing proportion of lower income households did not represent a resolution of the housing problem.

The difficulty of this approach is maintaining its credibility in the long run, especially when the policies pursued actually make the situation worse. The Conservatives' monetarist experiment, pursued in its purest form between 1979 and 1981, relied heavily on cuts in housing to achieve the desired reduction in total public expenditure. The Environment Committee calculated in 1980 that housing provided 75% or 92% (depending upon how the calculation was done) of the cuts in public expenditure planned in the period 1979-80 to 1983-84. (49) To offset the political impact of the housing cuts the Government placed increased emphasis on home ownership, but reliance on very high interest rates and readiness to tolerate record unemployment were parts of its economic policy which were in direct conflict with its housing policy. High interest rates merely made home ownership more expensive, especially for new and recent purchasers, and unemployment, or fear of unemployment, must have made mortgages unobtainable or unattractive for many people.

The housing crisis of the mid 1980s, therefore, exists on a number of different levels. It is a crisis for the millions of families and individuals who live in expensive or inadequate housing, or who have no home at all. It is a crisis for the system of housing provision that has developed over many years but which is now showing signs of severe strain. And it is a crisis of Government policy, which is increasingly seen to be contradictory: on the one hand home ownership is for the great majority, but economic policy makes this increasingly expensive, and on the other hand the role of the local authorities is seen in terms of provision for the least well off while resources are denied for further new building to meet their needs.

Conclusion

This chapter has sought to introduce the idea of the housing crisis in a fairly general way and in doing so has aimed at making three main points, which can be summarised as follows: i) that housing policy since the late 1960s has been characterised by growing complacency, ii) that long established trends in population change, the age of the housing stock and the pattern of housing tenure have contributed to a growing housing problem of crisis proportions in the mid 1980s; and iii) that the problem cannot be explained away in terms of the residual problems of special needs groups or pockets of urban residential decay, but must be seen as arising from structural features of the housing system and the wider economy.

Subsequent chapters deal with particular aspects of the present situation most of which have been mentioned above. Using a variety of approaches and data sources, they all share the same general perspective on the nature and severity of the housing crisis. First Philip Leather and Alan Murie look at the massive cuts in public sector housing investment and the various devices used by the Government, especially since 1979-80, to reduce both new building and repair and modernisation work by local authorities. They show, amongst other things, how capital receipts from the sale of council houses have become increasingly important in funding local authority capital expenditure, and how Treasury preoccupation with cash limits has overtaken housing policy considerations in deciding spending levels. This is followed by Ted Cantle's discussion of the very serious repair and maintenance problems which are emerging in the public sector. He argues that although the rapid deterioration of industrialised and system built housing of the 1960s and early 1970s has (quite rightly) received most attention, there is also an urgent need to invest in the pre-1945 stock. This discussion goes beyond detailing the problems of dampness, condensation, spalling concrete etc., to refer to the wider social consequences of neglect of public housing.

In chapter four Mike Gibson considers the evidence in the House Conditions Surveys which shows an increase in the number of houses in need of extensive repair, and concludes that the impending crisis is qualitatively and quantitatively different from past experience. This chapter goes on to look at the 1985 Green Paper on improvement policy and concludes that the prospects for the future are bleak, with not enough clear thinking being done now to plan for appropriate responses to the growing problem. The next five chapters look at the experiences of different groups of consumers and the policies that affect them. First Valerie Karn, John Doling and Bruce Stafford draw on their research to examine the problems of low income home owners. They show that low income purchasers are in many ways victims of home ownership rather than beneficiaries, in the sense that they live in the poorest quality housing, which they cannot afford to maintain or improve to a satisfactory standard, they pay a high proportion of income on mortgage repayments and obtain least assistance with their housing costs. It is argued that the policy of expanding home ownership is not accompanied by mechanisms for channelling help to those in greatest need. Next David Byrne looks at the impact of the recession and, in particular, changes in the level of unemployment locally, on the housing market and the position of existing home owners in that market. He shows that sudden steep increase in unemployment can severely disrupt the housing market, making it very difficult for existing owners to trade-up, thereby denying them access to one of the major advantages claimed for owner occupation and undermining the credibility of housing policy based so heavily on expansion of home ownership.

Christine Oldman looks at the elderly as a group amongst whom experience of poor housing conditions is particularly widespread. She argues that the main reason for poor housing among the elderly is the economic disadvantage that goes with old age, and yet unlike other disadvantaged groups, such as single parents there is a policy of providing high quality accommodation, sheltered housing, for a privileged section of elderly people. Sheltered housing is criticised for creating an elite and for diverting attention from the real problem of providing satisfactory standards in ordinary housing for the elderly.

Chris Holmes deals with a group who are virtually squeezed out of the housing system altogether, yet whose numbers have increased dramatically in recent years. Housing policy has never recognised the needs and aspirations of single people to form separate households, but it is argued here that recent developments in housing policy have greatly increased the numbers of single homeless. Chris Holmes charts the unplanned and very expensive growth of bed and breakfast accommodation since 1979, and argues that what is needed is fundamentally different policies to provide decent housing for single people. Peter Malpass also looks at a group, young adults, who have been given very low priority in housing policy in the past, unless they had dependent children of their own. Problems of access to decent housing at an affordable price are identified as the major issue for young people, and the Conservative Government's policy of further growth in home ownership is shown to be inappropriate to the needs of this age group.

The final chapter pulls together some of the themes emerging from the book as a whole, and considers the idea of a crisis as a turning point, questioning whether that point has yet been reached and how it can be reached in the prevailing political climate of the mid 1980s.

NOTES AND REFERENCES

(1) First Report of the Environment Committee, Session 1979-80, Enquiry into Implications of Government's Expenditure Plans 1980-81 to 1983-84 for the Housing Policies of the Deparment of the Environment. HC714, HMSO, London, 1980, P XVI.

(2) M Flemming and J Nellis, 'A New Housing Crises?' Lloyds Bank Review, April 1982.

(3) Association of Metropolitan Authorities, Housing in the Eighties, 1980; Ruin or Renewal, 1981; Building for Tomorrow, 1982.

(4) M Gibson and M Langstaff, 'Housing Renewal: Emerging Crisis and Prospects for the 1990s', Housing Review, September-October, 1984.

(5) M Ball, Housing Policy and Economic Power, Methuen, London, 1983, Ch. 1.

(6) S Lansley, in Labour Housing Group, Right to a Home, Spokesman, Nottingham, 1984.

(7) Inquiry into British Housing: The Evidence, NFHA, London, 1985.

(8) Shelter Election Briefing 1, 1983.

(9) Homeless Households Reported by Local Authorities in England, Department of the Environment, 1984.

(10) Housing and Construction Statistics, HMSO quarterly.

(11) Ibid.

(12) English House Conditions Survey 1981, Part 1 Report of the Physical Conditions Survey, HMSO, London, 1982.

(13) The Next Ten Years: Public Expenditure and Taxation into the 1990s, HMSO, London, 1984. p. 9.

(14) The Guardian, 16 March 1985.

(15) Housing Policy - A Consultative Document, Cmnd 6851, HMSO, 1977, p. 11.

(16) D Donnison and C Ungerson, Housing Policy, Penguin, Harmondsworth, 1982, p. 196.

(17) Ibid., p. 285.

(18) The Housing Programme 1965-70, Cmnd 2838, HMSO, 1965.

(19) Donnison and Ungerson, op. cit. p.287.

(20) See P Malpass and A Murie, Housing Policy and Practice, Macmillan, London, 1982, Ch. 3.
(21) F W S Craig, British General Election Manifestos 1900-1974, Macmillan, London, 1975, p. 288.
(22) M Gibson and M Langstaff, An Introduction to Urban Renewal, Hutchinson, London, 1982, p. 63.
(23) Old Houses into New Homes, Cmnd 3602, HMSO, 1968.
(24) A Crosland, 'The Finance of Housing', Housing Review, September-October 1975, pp. 128-30.
(25) Housing Policy - A Consultative Document op. cit. p. iv.
(26) Ibid., p. 44.
(27) M Clark, 'Too Much Housing?' Lloyds Bank Review, October 1977.
(28) S Lansley, Housing and Public Policy, Croom Helm, London, 1979, p. 218.
(29) M Harloe, The Green Paper on Housing Policy, in M Brown and S Baldwin, The Year Book of Social Policy in Britain 1977, Routledge and Kegan Paul, London, 1978.
(30) S Merrett, State Housing in Britain, Routledge and Kegan Paul, London, 1979, p. 269.
(31) M Ball, Housing Policy and Economic Power, Methuen, London, 1983, p. 3.
(32) S Merrett op. cit. p. 269.
(33) D Nevitt, British Housing Policy, Journal of Social Policy, Vol. 7, Part 3, July 1978, pp. 329-34.
(34) M Heseltine, Address to the Institute of Housing Annual Conference, Brighton, June, 1979, reprinted in Housing, September 1979.
(35) Ibid.
(36) First Report of the Environment Committee, op. cit. p. 6.
(37) See S Lansley, R Forrest and A Murie, A Foot on the Ladder?, SAUS, Working Paper No. 41. University of Bristol, 1984, J Littlewood and S Mason, Taking the Initiative, HMSO, London, 1984.
(38) 'Social Trends', 1983, HMSO, London, p. 13.
(39) Ibid., p. 24.
(40) J Ermisch, The Political Economy of Demographic Change, Heinemann, London, 1983, p. 192.
(41) Ibid., p. 185.
(42) Ibid., p. 186.
(43) Ibid., p. 187.
(44) Housing and Construction Statistics.
(45) Social Trends 1983, HMSO, London, 1982, p. 12.
(46) Annual Abstract of Statistics 1985, HMSO, London, 1984, Table 2.7.
(47) Housing and Construction Statistics 1970-

<u>1980</u>, HMSO, London, 1981, Table 109.
(48) P Malpass, 'Residualisation and the Restructuring of Housing Tenure', <u>Housing Review</u>, March-April 1983.
(49) <u>First Report of the Environment Committee</u>, op. cit. p. v.

Chapter Two

THE DECLINE IN PUBLIC EXPENDITURE

Philip Leather and Alan Murie

Introduction

One of the most notable and commented upon features of the Thatcher Government's housing policy is the severity of the cutbacks in public expenditure. The housing programme as set out in the annual Public Expenditure Survey Committee (PESC) White Papers containing the Government's spending plans has fallen in real terms from £6.6 billion in 1979/80 to a planned £2.1 billion in 1985/86 (at 1983/84 prices), (1) a reduction of 68%. Housing's share of total public spending has fallen from 7% in the mid-1970s to an anticipated 2% in 1987/88. The housing programme has borne the lion's share of public expenditure cutbacks overall and in 1980 the House of Commons Environment Committee noted that housing cutbacks accounted for 75% or more of all public spending reductions. (2) In 1983 O'Higgins noted:

> Housing is not only the welfare programme suffering the largest cuts, it is also the only programme where cuts have been overachieved: by 1981/82 the estimated outturn cut was one and a half times the size the government had planned in 1980. (3)

Although there was some overspending against cash limits in 1983/84 and 1984/85, this picture remains basically correct and future spending plans show no indication of any reversal in this trend. However, our purpose in this chapter is to argue that a categorisation of Government housing finance policy as one based solely on cuts is both over-simple and misleading. More detailed examination of spending trends and outcomes suggests that <u>the nature of the crisis</u> in housing public expenditure is not only extremely complex but also poses more intractable problems than can be solved by the reversal of cuts in existing spending programmes. We argue that on top of expenditure cuts there has also been a transfer in the balance of effective public expenditure away from the public sector and direct investment towards the support of owner occupation; and that much of this support has been based on the proceeds from the sale of public sector assets which in the longer term will have significant and adverse financial effects on the need for future investment.

What is 'Public Spending'?

The Government's (or Treasury's) view of what constitutes public spending is set out in the annual PESC White Papers already referred to. Table 1 summarises the main elements. We refer to this as the PESC housing programme, that is to say those elements of expenditure included in the 'Housing' Chapter of the annual Public Expenditure White Paper. However there are two additional and significant categories of public expenditure on housing which do not appear in this table: firstly items counted as public spending but incorporated in other programmes, of which the most important, Housing Benefit, appears in the Social Security programme; and secondly, items which are not defined as public spending at all, of which the most important include tax foregone as a result of mortgage interest relief, exemptions from capital gains tax of capital gains made on sale of main residence, and discounts to purchasers of council housing.

There is considerable debate as to whether these items should be counted as a subsidy to owner-occupiers and hence as public spending even though they are essentially income foregone rather than monies paid out. Economists including Ermisch (4) have argued that subsidies to both owner occupiers and renters should be defined with reference to the difference between market values and actual housing costs. (5) However, we take a more pragmatic approach here and refer to subsidies in terms of cashflow as amounts actually paid or in the case of tax relief and discounts, amounts foregone.

From this perspective, the sums involved in public spending on housing are somewhat different from the official picture, as Table 2 shows. In 1984/85 for example, while 3.1 billion is planned to be spent

Table 1: The PESC Housing Programme 1978/79-1987/88 (£m cash except where indicated)

				Actual				Estimated	Planned	
	78/79	79/80	80/81	81/82	82/83	83/84	84/85	85/86	86/87	87/88
Central Gov. Subsidies to LA Housing	1,004	1,274	1,423	906	507	281	360	400		
Rate Fund Contributions to LA Housing	200	321	430	420	434	504	475	296		
Other Current Expenditure	191	247	283	291	300	318	326	273		
Total Current Expenditure	1,395	1,842	2,136	1,617	1,241	1,103	1,161	969	950	920
LA Capital (Gross)	2,248	2,595	2,258	1,920	2,468	3,109	2,814	2,321		
(Constant 1983/84 Prices)		(3,774)	(2,766)	(2,139)	(2,576)	(3,109)	(2,680)	(2,127)		
LA Cap. Receipts	-501	-488	-568	-976	-1,739	-1,789	-1,465	-1,595		
LA Net Capital	1,747	2,146	1,691	944	729	1,320	1,349	726		
(Constant 1983/84 Prices)		(3,121)	(2,071)	(1,051)	(761)	(1,320)	(1,285)	(665)		
Housing Corp. Gross Cap.	327	401	508	522	755	734	687	685		
(Constant 1983/84 Prices)		(583)	(622)	(581)	(788)	(734)	(654)	(628)		
Housing Corp Cap Recpts	-3	-4	-13	-29	-95	-110	-74	-80		
Housing Corp. Net Capital	324	397	495	492	680	624	613	605		
(Constant 1983/84 Prices)		(577)	(606)	(548)	(710)	(624)	(584)	(555)		
Other Capital	105	137	142	79	12	5	-3	-10		
Total Capital	2,176	2,680	2,328	1,515	1,421	1,949	1,959	1,321	1,590	1,720
Total Programme	3,571	4,522	4,464	3,132	2,662	3,052	3,120	2,290	2,540	2,650
(Constant 1983/84 Prices)		(6,577)	(5,468)	(3,489)	(2,779)	(3,052)	(2,972)	(2,099)	(2,225)	(2,255)

Source: Cmnd 9428: The Government's Expenditure Plans 1985/86 to 1987/88 (HMSO 1985)

on the PESC housing programme, £2.5 billion will be spent on Housing Benefit, mortgage interest tax relief will cost some £3.5 billion and capital gains tax exemption some £2.5 billion. Public spending on mortgage tax relief was about a third of the level of spending on the PESC housing programme in 1979/80, but in 1984/85 it is expected to exceed it. High interest rates have contributed to this but unless restrictions are introduced tax relief will continue to rise as house prices increase. Over the same period spending on Housing Benefit has probably trebled and in 1985/6 it also is planned for the first time to exceed the PESC housing programme. Discount on sales in England and Wales between October 1980 and September 1983 amounted to £2.9 billion.

The categorisation of expenditure in Table 2 is of little policy significance and we look below in more detail at changes in the pattern of expenditure between tenures, between investment and other payments, between types of output, and between areas and types of local authority. Table 2 does however indicate that even if we confine our analysis to include Housing Benefit and mortgage interest tax relief in our definition of public spending, the total of expenditure after remaining fairly constant in cash terms has risen sharply since 1982/83 and even in real terms (applying the same GDP deflator assumptions to all types of spending) has only fallen overall by about 20% during the period in question. It is clear therefore that what we have seen is a major transfer of resources from those programmes conventionally included in the PESC housing programme (mainly subsidies to public sector housing and new capital investment) towards other expenditure (Housing Benefit and support to owner occupiers), rather than a simple cut in spending. In the Government's own terms this must count as a major failure in public expenditure policy; but on the other hand, the nature of the transfer, from the public sector to the support of owner occupation, from investment to subsidy, and from the subsidisation of the production of public housing to the subsidisation of individual consumption reflects other priorities which have tended to override expenditure considerations.

Investment and Subsidy

There has been concern about the balance of spending in housing between new investment and subsidy over a long period and much of the debate surrounding the 1977 Green Paper and the preceding review of Housing Finance focused on this issue in its broadest sense, including considerable attention given to the issue of subsidies to owner occupation and a number of detailed proposals for reform. (6)

The Conservative Government also affirmed its commitment:

> to reduce the overall level of housing subsidies over a period of years so as to enable a greater proportion of the resources available for public expenditure on housing to be devoted to capital rather than current expenditure. (7)

but as we have seen this concern was mainly focused on the narrow area of subsidies to local authority housing revenue accounts rather than subsidies generally. Although periodically signs of internal debate in Government circles about tax relief have emerged publicly, the Prime Minister has personally taken a strong line in defence of subsidies to owner occupiers and recently the maximum limit on eligibility for tax relief was raised following the general election of 1983.

Even the assault on general subsidy to local authority housing has not resulted in the reductions hoped for. The share of the total PESC housing programme accounted for by current expenditure (mainly subsidies to local authority and new town housing and administration costs) was 18% of the total in 1969/70 but rose steadily through the following decade to 45% in 1980/81 and 47% in the following year. By 1983/84 however it had fallen back to 36% and stayed at a similar level in 1984/85.

The rising subsidy bill in the early and mid-1970s reflected a combination of factors including rising investment levels and high interest rates. A new subsidy system (8) for local authority housing, designed specifically to give central government more control over the total subsidy bill was proposed in the 1977 Green Paper, but not introduced (under the Housing Act 1980) until 1981/82 by the Conservative Government. This system introduced a notional Housing Revenue Account calculation for each local authority with subsidy entitlement derived from assumptions determined by the Secretary of State for the Environment about increases in subsidisable expenditure and assumed income (mainly rents, but also contributions from the rate fund in a significant number of authorities).

Table 2: An Alternative Housing Programme (£m cash).

	1979/80	1980/81	1981/82	1982/83	1983/84	1984/85	1985/86	1986/87	1987/88
Public Sector(1) Gross Housing Capital Expenditure	3,152	2,928	2,558	3,298	3,923	3,578	3,051	3,120	3,150
Public Sector(1) Net Housing Capital Expenditure	2,680	2,328	1,515	1,421	1,949	1,959	1,321	1,590	1,720
PESC Housing Programme	4,522	4,464	3,132	2,662	3,052	3,120	2,290	2,540	2,650
Housing Benefit	932(2)	1,039(2)	1,395(2)	1,663(2)	2,133	2,547	2,690	2,800	
Tax Relief on Mortgage Interest(5)	1,639	2,188	2,292	2,456	2,767(3)	3,500	4,000(4)		
Exemption from Capital Gains Tax(5)			2,800	3,000	2,500	2,500			

Sources: Annual Public Expenditure White Papers 1979/80 to 1985/86; Hansard vol. 49 (25.11.83).

Notes:

(1) Local authorities, new towns, Housing Corporation and home loan scheme.
(2) Estimated figures. Payments of housing costs to supplementary benefit applicants not included in Housing Benefit for 1979-83.
(3) Excludes £55 million MIRAS to nontaxpayers.
(4) Estimated figure.
(5) Public Expenditure White Paper indicate that estimates to these items are subject to large margins or error and are affected by changes in the operation of this tax.

By requiring substantial increases in local contributions (rent or rates) (9) in the 1981-83 period the Government was able to reduce its own subsidy bill significantly over 1980/81 levels as Table 1 shows, although achieved reductions have consistently fallen short of those planned and subsidy is now planned to rise again gradually. Rate fund contributions to housing revenue accounts have also consistently exceeded Government plans by a large margin. Overall, the 1984/85 Public Expenditure White Paper shows that the expected percentage of current expenditure in 1984/85 (37%) contrasts sharply with the 'planned' figure in the previous White Paper which had indicated that current expenditure would form only 27% of the total in 1984/85.

One element in this is that rent increases have not continued at the pace of the early 1980s. The Goverment's assumed rent increase for subsidy purposes in 1985/86 for example was 60 pence per dwelling per week compared with £2.50 in 1982/83. Ministers had previously indicated an intention to maintain rent increases and had referred to a real rent increase of 10% in the period 1978/79 to 1983/84. (10) The outcome may reflect political and electoral considerations and the failure of controls applied via the Rate Support Grant system to achieve the Government's desired reductions in rate fund support for Housing Revenue accounts. It may also reflect the burden which rent increases imply for the social security budget (Housing Benefit payments) in a period of high unemployment. Estimates made for the Environment Select Committee in 1980 suggested that every £1 increase in rent resulted in 40p in additional housing benefit on average. (11) A proportion of the reduction in general subsidy to local authority housing might therefore represent a transfer to income-related Housing Benefit in the Social Security budget. Other factors include the continuing high level interest rates and the growing burden of disrepair in the Council Stock, especially in authorities still in receipt of subsidy.

Whatever the causes, the implications of this are that there is little scope for any increase in the proportion of spending within the overall PESC housing programme which is devoted to capital rather than current expenditure while the total programme remains so tightly constrained. In the Minister's own words "every penny that goes towards subsidies is a penny off housing investment". (12)

If we turn again to the wider picture of public investment shown in Table 2, it can be seen that the scale of net capital expenditure by the public sector in relation to other expenditure is very small indeed. Even if we exclude the receipts from the disposal of assets from this comparison (gross capital expenditure in Table 2) the picture is still dominated by subsidies, mainly to individuals via tax relief or Housing Benefit. If a tighter definition of investment is adopted (13) there may be periods when no net capital investment has occurred at all.

Subsidies per se are not necessarily undesirable. What is at issue is on the one hand the balance between new investment and support for individual households, and on the other the balance of support for households in different tenures. Without doubt the proportion of spending on new investment, net or gross, is far too small. Evidence of serious problems in the public and private housing stock and of the re-emergence of shortages of housing, is mounting. (14) The Government itself has recently expressed concern about the rate of replacement of inadequate housing. (15)

Despite this what we have seen since 1979 is a major increase in the level of support for owner occupiers through the increase in mortgage tax relief, council house sales discounts, and capital gains tax exemption. Even the rally in gross capital spending in 1983-85 was founded largely on increased expenditure on improvement grants to private owners. By contrast, a major drive has reduced public housing subsidies, but at the expense of increased costs in the payment and administration of income-related Housing Benefits. The issues of equity between tenures that this raises are starkly highlighted by the figures in Tables 1 and 2.

The Decline in Capital Expenditure

Table 1 shows the pattern of actual forecast or planned public sector capital expenditure over the 1979-86 period. Despite the governments's avowed concern at the level of the subsidies, the incoming Secretary of State Michael Heseltine turned his attention in 1979 to housing capital spending as the most immediate way of achieving rapid and substantial spending cuts. The previous Callaghan administration had also provided the framework for cutbacks by introducing cash limits on housing capital spending and the Housing Investment Programme (HIP) system of single-year authorisations for local authority expenditure. (16)

This approach proved so successful that it was soon extended to cover all local authority expenditure.

As a result of two successive cuts in the total authorised expenditure and a mid-year moratorium in 1980 on the letting of new contracts, gross local authority capital spending fell by 43% in real terms from 1979/80 to 1981/82. Housing Corporation-funded expenditure by Housing Associations by contrast fell hardly at all over the same period in real terms, and thus rose from under a quarter to about one third of the local authority level.

In 1982, partly under pressure from the construction industry but partly no doubt also as a result of the pending General Election the Government took measures to stimulate more capital spending. Local authorities were encouraged to let contracts and the level of certain improvement grants was increased for a two year period. The result was a considerable rallying of gross capital expenditure especially in the 1983-85 period although even at its peak gross spending remained well below 1979/80 levels. For 1985/86 however, it is again planned to reduce expenditure, in the case of local authorities, to something equivalent to the depths of the 1981/82 level in real terms.

This picture of fluctuations in gross capital expenditure must be qualified in several important ways. The results of the Right-to-Buy and other measures have been to produce an enormous increase in the level of capital receipts especially by local authorities. These rose four-fold, for example, between 1979/80 and 1982/83. As a result the level of net expenditure by local authorities has been significantly reduced, particularly since 1981. In real terms, net new capital expenditure by local authorities in 1982/83 was only about 25% of its 1979/80 level and in 1985/86 is planned to approach 20%. Although gross spending increased during 1983-85, the continuing high level of receipts reduced the rate of increase of net spending which even in 1983/84 only reached 40% of its 1979/80 level.

Capital Receipts

Clearly therefore, it is capital receipts which have supported the level of gross housing investment. The Government has used the level of receipts to permit the reduction of net new investment on a dramatic scale. In 1982/83 net capital expenditure by local authorities (£729 million) barely exceeded the net level of funding of housing associations by the Housing Corporation (£680 million). This raises major issues in relation to the use of capital assets: firstly the real value of assets disposed of is far higher as a result of discounts. Secondly, the disposal of assets will lead in the medium or long term to additional and greater expenditure (if any attempt is made to replace lost relets or meet demand) - for example on the acquisition of sites to replace those previously sold or the construction or acquisition of new housing. Finally, whatever view is taken of these longer-term arguments, and of the view that proceeds (however discounted) should be used for additional investment the prospect of declining receipts poses a real threat to the maintenance of even the reduced housing programme which capital receipts have facilitated.

Capital receipts associated with the housing programme have been more substantial than those from any other programme despite the publicity which some sales (such as British Telecom) have received. Sales of local authority dwellings have represented the most significant act of privatisation or sale of assets carried out in the period since 1979. In 1979/80 total receipts from special sales of assets were £370 million; in 1980/81 they were £405 million; in 1981/82 they were £494 million; in 1982/83 they were £488 million and in 1983/84 they were £1142 million. Total sales and repayments from the housing programme for these years were £448 million; £568 million; £976 million; £1739 million and £1789 million respectively.

Table 3 details the compositon of housing capital receipts since 1978/79. The volume of receipts has risen dramatically since 1979/80 with increasing levels of council house sales, and initial receipts from the sale of dwellings have supplanted repayments of loans as the principal source of receipts.

Council house sales in the period 1979-84 were more than double those in the previous forty years combined and have exceeded those completed in the whole history of council housing. Neither sales nor capital receipts are a totally new phenomenon. The main difference in recent times however is in the relationship of the volume of capital receipts to capital expenditure. The local authority housing capital programme has become substantially self-financing. Proceeds from the sale of assets have become the major source of investment funds and the call of housing on the public sector borrowing requirement

Table 3: Sources of Local Authority Housing Capital receipts (£ million cash)

	1978-79		1979-80		1980-81		1981-82		1982-83		1983-84		1984-85	
	%	£m	%	£m	%	£m	%	£m	%	£m	%	£m	%	£m
Sales of land and other assets	24	5	38	8	96	17	99	10	135	8	103	6	105	7
Initial receipts from sales of dwellings	138	28	122	27	186	33	532	55	1,017	58	970	54	830	57
Repayments of sums outstanding on sales of dwellings	40	8	43	10	53	9	89	9	282	16	445	25	315	22
Repayments of loans to private persons	293	58	241	54	216	38	240	25	287	17	246	14	185	13
Repayment of loans to Housing Associations	7	1	4	1	16	3	15	2	18	1	25	1	30	2
Total	501	100	448	100	568	100	976	100	1,739	100	1,789	100	1,465	100

Source: Cmnd 9143 II and 9428 II and Hansard vol. 49, 23.11.83, cols. 212-3

has been considerably reduced as a result.

The role of the building societies in this has been crucial. Their involvement in financing council house sales has enabled initial receipts from sales to rise well above the historic levels associated with sales in the 1970s.

During 1982 building societies financed approximately 75,000 purchases of council houses by sitting tenants. This was equivalent to about 9% of all building society loans for house purchase and 16% of those to first time buyers. There was a considerable increase through the year in the number and size of mortgage advances and consequently, in the amount of lending. The 5% sample survey of building society mortgage completions indicates total lending of £682 million in 1982.

Thus, while the capital value of sales over the period 1979-83 levelled out at just below £500 million a quarter, the proportion of this capital value yielding an immediate capital receipt rose dramatically. At the end of 1979 and in 1980, prior to the operation of the Right to Buy initial payments represented 30% or less of capital value. However there has been an increase since mid 1980 to a peak of 67% at the end of the period under discussion largely as a result of building society involvement. Hence the sale of dwellings in the first quarter of 1983 yielded 56% more in initial payments than the larger number of dwellings sold in the first quarter of 1982.

The involvement is only partly within the capacity of government to control. The flow of funds into and out of building societies is influenced by a range of factors and the availability of funds for council house purchase will also depend on other demands being made. However building societies appear to have adopted the view that they should meet the demand for borrowing for house purchase. They have achieved this by raising interest rates to the level needed to attract enough investment. In this sense large capital receipts associated with building society funding of council house sales have involved higher interest rates for all house purchasers with a mortgage than would otherwise have been necessary.

A number of arguments could be advanced over whether capital receipts 'should' be reinvested in housing. Without entering into such a debate at length it is apparent that the housing sector's demands on 'non-housing' funds have been more substantially cut than is compatible with the level of gross capital expenditure. Without housing capital receipts in recent years either the housing programme would have been even more severely cut or housing would have come into sharper conflict with other expenditure programmes in resource demands. Table 4 demonstrates the divergence between net and gross expenditure (in cash terms) and the particularly striking divergence for local authorities.

Table 4: Index of Gross and Net Housing Capital Expenditure 1978/79 to 1984/85

(1978/79 = 100)

	Total Expenditure		Local Authority Expenditure	
	Gross	Net	Gross	Net
1978-79	100	100	100	100
1979-80	111	153	115	123
1980-81	108	133	100	97
1981-82	95	87	85	54
1982-83	122	81	110	42
1983-84	145	112	138	76
1984-85	132	112	125	77
1985-86	113	76	103	42

Source: calculated from Cmnd 9428.

This dependency on receipts also poses problems for the future. Forecasting capital receipts on an annual basis is not easy. It is not just a case of anticipating the level of sales activity. Even more problematic are estimates of the capital value of sales, the levels of discounts applied and the proportion of sale price being received in initial payments. The assumptions about council house sales on which public and capital expenditure are based are all highly problematic assumptions. The dependency of the housing programme on receipts consequently involves a potentially wasteful fluctuation and instability as well as a substitution problem if receipts decline.

The impending decline of receipts has threatened the housing programme for several years without yet materialising. In part it is in the Government's interest to underestimate the eventual level of receipts in a given year since any extra receipts reduce net spending, provided gross spending does not exceed the original target. However, it seems likely that without new measures to stimulate them, the main source of housing receipts will decline in future, particularly as the demand for council house purchase by sitting tenants falls off. However the crucial factor, the rate of decline, is more difficult to pin down. Finally, it is also possible that other sources of receipts such as large-scale sales of local authority estates to private or quasi-public organisations funded by the private sector, may become more significant. The accumulated backlog of receipts could also serve as a cushion for some years. Any significant falling off in receipts in the long term will however require substantial additional new money or a further significant cutback in the capital programme.

Local Authority Capital Programme

In addition to their overall significance in relation to capital spending, capital receipts have also become of greater importance in determining local levels of spending since the introduction of a new system of capital expenditure controls under the Local Government, Planning and Land Act, 1980.

Prior to 1981 reinvestment of capital receipts was only possible to the extent explicitly allowed through loan sanction. HIP allocations provided a ceiling for borrowing for housing capital expenditure and capital receipts could not be used to augment this amount. As housing debt is not earmarked, the choice for treasurers was between writing off HRA debt and thereby reducing loan charges or making interest payments to the HRA in respect of accumulating capital receipts.

Under the arrangements introduced from 1981/82 individual local authorities are able to increase their block allocations for capital spending for all services by a proportion of the housing capital receipts which they receive in the course of the year and a proportion of accumulated receipts. The remaining 50%, as we have seen, is taken into account at the national level and its effects are distributed in the HIP allocations for the year in the normal way. To the extent that an authority does not make use of receipts in the year in which they accrue, it can carry them forward as a basis for spending over and above its allocation in a subsequent year.

The scope for supplementing allocations with additional resources has become more and more important each year since 1981/82 (see Table 6) and it is worth considering how it works, in some detail. Having estimated the total of capital receipts for the coming year, the Government then adds this to the total additional borrowing for the year which it wishes to see, to arrive at a level of 'gross provision'. For 1984/85 this was £2,522 million. The Government then estimates the proportion of capital receipts which, under the existing rules of the system, local authorities may use themselves to supplement their allocations (called the 'prescribed proportion'). This is deducted from gross provision and the remaining sum is available for HIP allocation.

Table 5: Capital Receipts and HIP Allocations 1984/85

	£m
Gross Provision	2,522
less Estimated Capital Receipts	1,465
equals Net Provision	1,028
Prescribed proportion of estimated capital receipts	666
Remaining receipts	799
adding Remaining receipts to Net Provision plus tolerance from 1982/83 gives:	
Total Available for HIP Allocations	1,853

Table 6 shows how the post 1981/82 system has influenced the sums available for allocation. From 1981/82 to 1983/84 a steadily increasing proportion of the total available for local authorities to spend was not distributed through allocations but permitted to accrue to local authorities on the basis of their capital receipts.

For 1981/82 and 1982/83 local authorities collectively underspent the resources potentially available to them (although most spent their HIP allocation), but in 1983/84 and 1984/85 there was major overspending against the Government's cash limit, although no individual authority overspent under the rules of the system. This is because local authorities began to spend the accumulated receipts from 1981/82, 1982/83 and previous years, which the rules of the system permitted them to use. These had built up to an estiamted £5 billion by the end of 1984. Given the difficulties in forecasting levels of receipts it is not surprising that many authorities waited until receipts were generated before planning to spend them. However the Government has reacted to this curious phenomenon of a 'collective overspend' by changing the rules and placing tighter restrictions on the prescribed proportion of receipts. For 1985/86 for example authorities can now use only 20% of accumulated receipts from the sale of council houses, compared to 40% for 1984/85 and 50% prior to this. At the time of writing a further review is in progress which is expected to impose tighter restrictions, especially on accumulated receipts from past years. The impact of this system on capital spending programmes has been severe, and an increasing volume of informed criticism is emerging, including some from unexpected quarters. (17) Local authorities have found a plethora of rule-changes, cutting their entitlement to spend at a stroke, moratoria on spending, and even basic confusion about the interpretation of legislation on the part of DOE itself. (18) These have induced a climate of uncertainty which in some circumstances has led to wastage or hurried decisions on major projects, to the winding down of programmes, staff redundancies, and other problems. Positive proposals for reform have included a longer time scale for spending plans (with three to five year allocations replacing the present single year allocations) and the reintroduction of longer-term strategies for the direction of investment priorities. The present system clearly provides neither the control which the Government has sought, nor the longer-term stability which effective capital spending programmes require.

Changes in the Pattern of Investment

There have also been important changes in the pattern of investment under the Conservatives, although some of these were already evident in 1979. It should be

Table 6: HIP Allocations and Expenditure

£000s

	1978/79	1979/80	1980/81	1981/82	1982/83	1983/84	1984/85
HIP Allocation	2424	2862	2186	1786	1875	1801	1853
Additional Capital Receipts available				413	493	684	666
Outturn HIP Expenditure	2249	2595	2258	1920	2468	3109	
Actual Capital Receipts	515	448	568	976	1739	1789	
Net Capital Expenditure	1734	2135	1675	982	668	1320	

Sources: HIP Allocation letters (various years) and Cmnd 9428 The Government's Expenditure Plans 1985/86 to 1987/88, (HMSO 1985).

borne in mind that in a context of overall decline in investment, a contracting share of the total represents an accelerated reduction in absolute terms. Table 7 shows the most significant developments in shares of gross capital spending within the local authority capital programme and between the three types of spending organisations - local authorities, new towns, and housing associations. Looking firstly at shares of total capital spending, the striking feature is the steady rise in the share of gross spending accounted for by the Housing Corporation's direct funding of Housing Associations. (It should be noted that local authorities also fund associations although the proportion of their resources devoted to this has gradually declined since its peak in 1977/78). This rise was only interrupted in the 1983/85 period by additional local authority expenditure based on the use of their capital receipts, and it is planned to resume in 1985/86. By contrast expenditure by new towns has virtually ceased even in gross terms.

The shift of resources towards housing associations is even more evident if expenditure net of capital receipts is examined. The scope for the generation of capital receipts by associations is limited as the majority of co-ownership housing built by them in the 60s and early 70s has now been disposed of. As a result the Housing Corporation programme takes a much larger share of net capital expenditure than Table 7 would suggest. In 1979/80 the Corporation's programme accounted for 15% of net spending; by 1982/83 this had risen to 48%; and in 1985/86 the figure is planned to remain high at 46% with the new towns actually planned to generate more in receipts than they invest during that year.

The significance of this expanded role for the Housing Corporation is also considerably emphasised if we examine output figures, particularly for the construction of new dwellings. As Table 7 shows, the percentage of local authority investment going to new construction has fallen from a peak of 60% in the mid-1970s to around a quarter at the present time. Housing Associations are providing about one third of all public sector new dwelling completions at the present time (Table 8) and an even higher proportion of starts in the first half of the financial year 1984/85. The information on public sector renovations is too unspecific to provide evidence but we would also suggest that housing associations are now carrying out the majority of public rehabilitation investment in the pre-1919 acquired housing stock.

Table 7: The Changing Pattern of Housing Capital Spending 1974-86

Programme	74/75	75/76	76/77	77/78	78/79	79/80	80/81	81/82	82/83	83/84(2)	84/85(3)	85/86
LA Spending (As % of LA Programme):												
New Building and Land	41	54	60	60	51	43	45	39	30	24(1)	25(1)	NA
Slum Clearance	3	3	3	2	3	3	5	5	3	1	1	NA
Renovation of Local Authority Stock	13	12	14	17	23	29	31	34	40	36	36	NA
Acquisition by Local Authority	8	5	4	3	3	2	1	1	1	-	-	NA
Improvement Grants	5	3	3	3	4	5	6	11	19	30	28	NA
Mortgage Lending	25	16	8	6	8	10	5	2	2	1	1	NA
Loans to Housing Association	5	7	9	10	8	7	8	7	6	4	5	NA
(As % of Total Programme)												
LA Total (Incl. Lending to Housing Associations)	91	88	85	84	83	82	77	75	75	79	79	76
New Towns	5	5	6	6	5	5	6	4	2	2	2	1
Housing Corporation	4(4)	6(4)	9(4)	11(4)	12	13	17	20	23	19	19	22

Figures may not add to 100 because of rounding.

Source: Leather, P., 'Housing Need and Housing Investment', Housing Review Nov-Dec 1983 and Cmnd 9428 (HMSO 1985)

Notes:
(1) Includes acquisitions by local authorities. Reclassifications in Cmnd 9428 do not permit separation.
(2) Reclassifications in Cmnd 9428 distinguish Enveloping and Environment Improvements (2%) previously included in LA renovations; and Low Cost Home Ownership schemes (1%) previously under various headings.
(3) Estimated outturn from Cmnd 9428. Also Enveloping etc (3%) and Low Cost Home Ownership (1%).
(4) Housing Corporation Net Expenditure.

Table 8: Permanent New Dwellings Completed: Public sector: England

	Number of Dwellings			
	Local Authorities	New Towns	Housing Associations	Total Public (1)
1979-80	70,000	7,000	16,700	94,300
1980-81	65,400	7,700	19,700	93,300
1981-82	39,700	7,900	14,800	62,400
1982-83	26,900	2,300	9,700	38,900
1983-84	28,200	1,100	14,000	43,300
1984-85 first six months (provisional)	13,900	1,000	6,400	21,300

1 - Including government departments.

Source: Cmnd 9438.

Where then is local authority investment being channelled? Clearly the two areas of relative growth are in renovations to local authorities' own housing stock and improvement grants (Table 7). Since 1983 these two programmes have accounted for over two thirds of local authority capital investment.

The majority of investment in the renovation of local authority stock relates to the purpose built council stock. The demand for investment has mushroomed in this area not only as a result of government initiatives such as the Housing Defects Act, the Priority Estate Projects, and the recently-established Urban Renewal Unit but also as awareness of the scale of problems in the traditionally-built stock, the 'non-traditional' stock of the 50s and 60s and the industralised and system built dwellings of the 1960s and 1970s has grown. (19) In many authorities the condition of the council-owned stock has become a political issue of equal or greater significance than more traditional inner city problems and the condition of the older private sector stock. Despite the growth in expenditure, the backlog of work is enormous.

The other area of expansion is in the improvement grants to private owners. Encouraged by higher rates of grant in the 1982-84 period the volume of applications and the level of uptake of grants (especially repair grants) has increased and expenditure rose seven fold between 1979/80 and the peak year of 1983/84, without taking into account related spending on enveloping schemes and environmental improvements in declared improvement areas.

A number of comments can be made on this expansion. Firstly, it is clear that the boost in expenditure based on the use of capital receipts was largely expenditure on renovation grants. Remaining programmes (including renovations to authorities' own stock) remained constant or declined. The rally in capital expenditure referred to was therefore primarily expenditure on the owner occupied stock. Subsequently the Government has suggested that much of this investment might have taken place without grants and has made proposals for a major scaling down (or 'targeting') of improvement grants. (20) Here then we have another example of an increase in financial support for the owner occupied sector, this time on the capital side. Again one might question the wisdom or the equity of using the proceeds from council house sales to finance private sector improvement grants.

The Regional and Local Allocation of Resources

There have also been significant shifts in the distribution of resources between different regions and localities. Table 9 shows changing shares of capital expenditure by region and by type of local authority over the past 10 years. The three northern regions boosted their share of expenditure considerably in the late 1970s but this trend has now been halted. London's expenditure share has by contrast fallen, slowly at first but more steeply in the last two years. The brunt of this decline has been borne by the GLC. The complement of this has been an increase in the share of spending in the South East, South West, Eastern and East Midlands regions.

The metropolitan districts collectively increased their share of spending in every year from their creation in 1974 until 1980, but this has now also fallen back to a lower level. Correspondingly, the non-metropolitan districts have pushed up their share of spending by more than eight percentage points over the last two years after a similarly long and steady decline. The pattern which seems to emerge overall from the table is that of a significant shift in the last two years which is sharper than any for many years previously. The beneficiaries are the shire districts in the Midlands, South and East, while the main losers are the metropolitan districts and London. In the north the picture overall is static but there has been a small shift from the metropolitan to the shire districts.

Some clues as to the causes of these changes are provided by Table 10 which shows shares of allocations and prescribed capital receipts. The Final Basic allocations in the table are initial HIP allocations as modified by various tolerance and supplementary allocations during, or even as in 1982/83 after, the spending year. Capital Receipts are the prescribed proportion of housing capital receipts up to which authorities may increase their spending over and above allocations. Permitted Spend is the sum of the Final Basic allocation and Capital Receipts. It is not, it should be emphasised, the total spending which authorities may undertake as they may have non-housing receipts from previous years. Initial allocations are based on a combination of an index of need and DOE Regional Office discretion. They should therefore bear some relation to the pattern of housing need. Apart from accumulated capital receipts they are also the only indication of available resources that authorities have prior to the spending year (although only three to four months in advance) and must exercise a strong influence on the scale of the programme which authorities plan for at that crucial stage.

Prior to 1981/82, expenditure shares bore a fairly close relationship to Initial allocations, but in 1981/82 and subsequently this relationship broke down. Up to 1980/81, allocations produced a gradual increase in the share of resources for the northern regions, and the metropolitan districts, at the expense of the Midlands, East and South, with London's share remaining fairly stable. In 1981/82, however, London's share of the Initial allocation was substantially reduced as a result of changes in the method of determining allocations and the data sources used. These resources were distributed fairly evenly across the regions of the rest of the country. In the following year London suffered a further but smaller reduction, which was re-allocated to the northern regions. Finally in 1983/84 this trend was reversed and the northern regions have subsequently lost resources to the south.

Changes in the Initial allocations do not by any means fully account for the shift in investment patterns. Table 10 shows clearly that for both 1981/82 and 1982/83 the pattern of Final Basic allocations

Table 9: Shares of Capital Expenditure by Region and Type of Authority, 1974-83

	Percentage of National Expenditure for Year								
	74/75	75/76	76/77	77/78	78/79	79/80	80/81	81/82	82/83
North (1)	27	26	25	25	27	30	30	29	30
Midlands, East, South East & South West (2)	39	41	41	39	38	37	35	38	41
London	34	35	34	36	35	34	35	32	29
Inner London	12	14	16	17	17	17	19	17	16
Outer London	11	10	9	8	8	8	9	9	8
GLC	11	10	10	11	10	9	7	6	5
Shire Districts	43	42	43	41	40	39	38	43	46
Metropolitan Districts	22	23	23	23	25	27	27	25	26

1 - DOE regions North, North West, and Yorkshire & Humberside.

2 - DOE regions East & West Midlands, East, South East and South West.

Table 10: Shares of Final Basic (1) HIP Allocations, Prescribed Capital Receipts, (2) and Permitted Spend, (3) by Region and Type of Authority, 1981-83.

Area	Percentage of National Total for Year 1981/82			1982/83		
	Final Basic	Capital Receipts	Permitted Spend	Final Basic	Capital Receipts	Permitted Spend
North (4)	30	23	29	32	24	29
Midlands, East, South East and South West (4)	39	49	41	40	55	44
London	31	28	30	28	21	26
Inner London	17	10	16	17	7	14
Outer London	9	9	9	8	9	8
GLC	5	9	6	3	5	4
Shire Districts	43	55	46	44	59	49
Metropolitan Districts	26	17	24	27	21	25

1,2,3 - For explanations, see text.

4 - For composition of regions, see Table 9.

and, even more so, Permitted Spend, was very different from that of the Initial allocations. In particular, the shire districts and the south (excluding London) were able to increase their share of resources. The reasons for this are also clear from the table: the Capital Receipts generated have been greater in these authorities than in the northern regions, the metropolitan districts, and inner London, and therefore the shares of Permitted Spend available to different areas are changed considerably when compared to the distribution of Initial allocations. The shire districts, for example, had 58% of the potential to spend generated by Capital Receipts, and hence over 48% of permitted spend, in 1982/83 compared to 42% of the Initial allocations. This represents a very substantial distortion of the original pattern.

There are two points to be made here. Firstly, the allocation process which has determined these swings in resources deserves greater attention than it has received in the past. Briefly, DOE uses a series of measures of housing needs, weighted and combined into a single index, the Generalised Needs Index (GNI) to determine allocations for housing capital spending under the HIP system.

The GNI measures a number of separate readily quantifiable housing problems which are weighted and combined to produce a relative measure of the needs of each authority or region compared to others. Allocations are firstly made to regions, corresponding to the areas administered by DOE regional offices, on the basis of their shares of the total GNI score. Cost compensation is applied to reflect inter-regional cost differences, and finally damping is applied to limit shifts in regional shares of allocation from one year to another. Regional allocations are then distributed amongst local authorities, partly on the basis of local GNI scores, and partly at the discretion of regional offices.

In practice, at local level discretion has to date been used very selectively in order to match the allocations suggested by GNI scores more closely to commitments and to past spending patterns. As a result, local shares of resources since the introduction of HIPs have changed remarkably little over time. 1983/84 Initial allocations, for example, are more closely correlated to 1975/76 expenditure than to 1983/84 local GNI scores. But at regional level, GNI scores are much more influential in shifting resources and the effects work through to the local level via regional control totals. For this reason the GNI is of considerable importance even to individual authorities.

The annual changes in regional GNI shares suggest strongly that it is not an independent or 'objective' measure, but one which has been subject to constant amendment as a result of political pressures. The year-to-year shifts in resources which it produces are considerable, even after damping. At local level these changes can be even more dramatic but cannot realistically reflect actual changes in housing problems. Yet apart from a brief annual consultation at the Housing Consultative Council, the composition of the GNI and the way in which it is used are centrally determined and little discussed.

Secondly, the system whereby the actual pattern of generation of capital receipts is used as a substitute for the allocation of spending permissions on the basis of need is producing a distorted pattern of spending. Council house sales which, as has been shown, are the largest item in the generation of capital receipts are highest in shire districts where owner occupation is already high. (21) The highest levels of spending permissions are accruing to those authorities which have least housing need, and often also least capacity to spend. For many, the level of receipts is very high in relation to their initial allocation and their recent expenditure levels. About 100 authorities had prescribed capital receipts of 75% or more of their Final Allocation in 1982/83. Even without pressures to limit revenue spending, it would be unreasonable to expect these authorities to increase their programmes on the scale required to spend the resources available to them through capital receipts. To take an extreme example, one authority had an allocation for 1983/84 of £2.9 million and prescribed housing capital receipts of £7.8 million.

The solution to this problem would be to return to the pre-1981/82 arrangements whereby the estimated national total of capital receipts was taken into account when setting the cash limit and its 'benefit' was effectively allocated to all authorities. Unfortunately the Government has consistently used reductions in the prescribed proportion of receipts 'netted off' under the present system to <u>reduce</u> not supplement the basic cash limit.

Finally, it is worth noting that debates about the regional distribution of resources have recently come to the fore in relation to Housing Corporation allocations to Housing Associations. The use of a similar GNI has resulted in pressures to shift resources from north to south. Whether any of these changes reflect a conscious intention to reduce levels in the north and in Inner London is difficult to say conclusively - we can only observe that significant spatial shifts of resources are occurring even within the reduced overall level of spending.

Impact on the HRA

Established texts referring to Housing Revenue Account Income make little or no reference to interest on capital receipts. However the 'underspending' of capital receipts is contributing to other changes in council housing finance. The sums which are available to finance housing capital expenditure (including accumulated receipts) or which are not available for this purpose can be lent to the authority's central loans pool, used to pay off appropriate debt, or used (instead of new borrowing) to finance non-housing capital expenditure. Table 11 indicates that local authority Housing Revenue Accounts have been significantly affected by this. By 1983/84 the estimated income from interest on capital receipts from council house sales was as important as rate fund contributions - including mandatory contributions associated with the administration of rent rebates. The combined income from interest on sales and other non-rent income exceeded exchequer subsidies and was growing in importance relative to rents.

Table 12 indicates that interest on council house sales has grown more rapidly than other items of HRA income since 1981. The 1983/84 figure for interest on council house sales (£494 million) compares with £81 million (2% of HRA income) in 1979/80. The recent increase in this income from interest has varied considerably but it is substantial in all the major categories of housing authority.

The importance of interest on council house sales for HRA is as another source of income which has enabled some local authorities to limit rent increases or rate fund contributions. However, as the majority of local authorities have moved into surplus on their HRAs a number of options have emerged. Some local authorities have established substantial Housing Repairs Accounts and some have made net transfers from their HRAs to their general rate funds. With the repeal of the previous restrictions on working balances in the HRA under the Housing Act 1980, 18 local authorities made such transfers in 1980/81. In 1981/82, 70 made transfers, in 1982/83, 78 and in

Table 11: HRA Income All Authorities England and Wales

	1977-78 (1)	1978-79 (1)	1979-80 (1)	1980-81 (1)	1981-82 (2)	1982-83 (2)	1983-84 (2)
Net Rent Income	42%	40%	37%	37%	54%	55%	55%
Government Subsidies	43%	43%	44%	39%	26%	20%	16%
Rate Fund Contribution	9%	10%	12%	13%	10%	10%	10%
Other Income; Interest on Sale of Council Houses	6%	7%	7%	11%	(5% (5%	6% 9%	9% 10%

1 - Actuals

2 - Estimates

Source: Housing Revenue Account Statistics, CIPFA
1978-79 Actuals; 1979-80 Actuals; 1980-81 Actuals; 1982-85 Estimates; 1983-84 Estimates.

Table 12: Components of HRA Income

	1981-82 £000s	1982-83 £000s	1983-84 £000s	Change 81-82 to 82-83 %	Change 82-83 to 83-84 %	Change 81-82 to 83-84 %
Net Rent Income	2,678,173	2,965,036	2,952,763	+10.7	-0.4	+10.3
Government Subsidies	1,292,214	1,086,699	873,433	-15.9	-19.6	-32.4
RF Contributions (Inc. Admin. of Rebates)	490,743	537,470	548,734	+9.5	+2.1	+11.8
Interest on Council House Sales	269,837	455,264	494,191	+68.7	+8.6	+83.1
Other Income	290,082	302,922	469,263	+4.5	+54.9	+61.8
Total Income	5,021,049	5,347,461	5,338,384	+6.5	-0.2	+6.3

Source: HRA Statistics 1982-83 Estimates; 1983-84 Estimates, CIPFA.

1983/84, 88 local authorities estimated that they would make such transfers. (22) In 1984-85, 97 authorities made such transfers. (23)

However, rather than take this course of action it appears that some local authorities in 1983/84 have chosen to restrict rent increases, ignoring the Secretary of State's determinations and leaving their HRA out of line with the notional HRA on which their subsidy is calculated. Of the 367 local housing authorities in England, 152 had reported no increase in rents and a further 14 had reported reductions in rents. This raises two immediate concerns. Firstly, in the Governments's terms there is a direct link between rents and capital investment levels and this phenomenon represents a failure to channel resources to investment or achieve reasonably equitable rent levels between different parts of the country. Secondly, and more importantly, the increasing gap between notional accounts on which subsidy is calculated and actual HRAs is likely to increase the reluctance of the authorities affected in this way to invest in the future. Initially it means debt associated with new investment would not benefit from exchequer subsidy and would have to be wholly funded by rent or rate increases. The operation of the housing subsidy system and other factors are thus providing a recipe for low investment rather than a pattern of activity either related to need or to what the country can afford.

Conclusions

How do cuts in public expenditure and even a reorientation in the pattern of spending represent a crisis in housing finance? And if the pattern could be easily and quickly reversed would problems still remain? Most of the discussion of cuts and expenditure changes in recent years is essentially arguing that current levels of investment and activity are insufficient and that problems are being stored up for the future. The private sector has not responded to the situation by substantially increasing investment. What stands out in particular is the scale of these problems. Later chapters in this book draw attention both to the scale and costs of renewal work required in the public sector stock and the rehabilitation/clearance issue in the private sector. In retrospect the levels of housing investment achieved in Britain since the mid-1970s will appear unreasonably low and the complacency which has surrounded the debate about expenditure and which has characterised the strategy of successive Ministers will be difficult to understand. It is not easy to develop a programme to 'catch up' on lost years of investment and any attempts to catch up raises doubts about the costs and organisation of production. In this sense there is a crisis rather than merely a period of low expenditure which is easily remedied subsequently. The crisis is compounded by other factors. The subsidy system has not been rationalised and new anomalies have emerged which are likely to inhibit local authorities in investment decisions. If government sought to use local authorities to spearhead an expansion of housing investment new subsidies and incentives would seem necessary. The burden of funding owner occupation will also not prove easy to throw off. Nor will the need for income support to meet rents. The costs of housing benefit and various supports to owner occupation are growing rather than declining. As capital receipts decline the competing demands for resources in housing will become more apparent and if the public expenditure orthodoxy is maintained will prove impossible to meet. The emerging crisis will be a public expenditure one raising questions about the economic case for maintaining controls over housing investment. It will also raise questions about cost and the organisation of production as well as the rationalisation of the whole structure of housing finance and subsidy. The crisis is one of diminishing room to manoeuvre alongside increasing need for innovation and change. It is not one which is likely to be resolved within the existing framework of housing public expenditure or of housing legislation and policy.

NOTES AND REFERENCES

(1) Most figures in this chapter, except where otherwise indicated, are given in cash (or actual expenditure) terms, and thus take no account of the tendency of inflation to erode cash expenditure over a period of time, particularly in relation to capital expenditure or more specifically, investment. Although it is difficult to produce firm figures as a result of variations in cost increases between and within spending leads and shifts in the composition of spending, the Government's own overall measure, known as the GDP deflator, gives a crude indication of these charges. It should however be borne in mind when talking about capital expenditure that the re-

ductions in output associated with a given level of lost investment are likely to be greater than suggested by an overall measure such as this.

(2) House of Commons Environment Committee, First Report from the Environment Committee Session 1979-80, Enquiry into Implications of the Government's Expenditure Plans 1980-81 to 1983-84 for the Housing Policies of the Department of the Environment, HC 714, HMSO 1980

(3) M. O'Higgins, 'Rolling Back to the Welfare State: The Rhetoric and Reality of Public Expenditure and Social Policy under the Conservative Government', in C. Jones and J. Stevenson (eds.) The Year Book of Social Policy in Britain 1982, Routledge and Kegan Paul 1983

(4) J. Ermisch, Housing Finance, PEP 198

(5) This viewpoint is well set out in an unsigned article 'Housing Finance and Subsidies: How to Devise a Policy Framework', Public Money June 1982

(6) Stewart Lansley gives a detailed account of these issues in Housing and Public Policy, Croom Helm 1979

(7) Cmnd 8175, The Government's Expenditure Plans 1981/82 to 1983/84, HMSO 1981

(8) See Bramley G., Leather P., and Hill M.J., Developments in Housing Finance, SAUS Working Paper 24, University of Bristol 1981 for a detailed description of the system

(9) Under the 1980 Act subsidy system the Secretary of State (after consultation) can set a level or levels of assumed increase in local income to the Housing Revenue Account. The larger the assumed increase the less the entitlement to subsidy. The advantage of this system to the Government is that it enables the approximate annual commitment to subsidy payment to be determined in advance

(10) House of Commons Environment Committee, op. cit. p. xi

(11) Ibid., p. xii

(12) House of Commons Hansard, vol. 28, 21.7.82, cols. 386-87 (Sir George Younger)

(13) See for example 'Crisis in Construction', Labour Research, July 1983, vol. 72 no. 7

(14) See for example the final report of the Inquiry into British Housing (chaired by the Duke of Edinburgh)

(15) CMND 9513, Home Improvement - A New Approach, HMSO, 1985

(16) See Leather P., 'Housing (Dis?) Investment Programmes', Policy and Politics 11(2), 1983 pp. 215-229, for a fuller discussion of the impact of HIPs

on capital spending.

(17) Capital Expenditure Controls in Local Government in England: a report by the Audit Commission, HMSO 1985

(18) Particular confusion reigned over the rules for tolerance (carry forward past or anticipating future allocations)

(19) The Association of Metropolitan Authorities has published three reports outling the nature of problems: Defects in Housing 1: Non-traditional Dwellings of the 1940s and 1950s, (July 1983), Defects in Housing 2: Industrialised and System-built Dwellings of the 1960s and 1970s, (March 1984), Defects in Housing 3: Repair and Modernisation of Traditional Built Dwellings (March 1985)

(20) Cmnd 9513, Home Improvement - A New Approach, HMAO 1985

(21) See R. Forrest and A. Murie, Monitoring the Right to Buy 1980-82, Working Paper no. 40, School for Advanced Urban Studies, University of Bristol 1984 and R. Forrest and A. Murie, Monitoring the Right to Buy 1980-85, School for Advanced Urban Studies, University of Bristol 1985

(22) House of Commons, Hansard, 3 Feb 1984, col. 372

(23) P. Malpass, Beyond Deterrent Rents, Paper presented to the ESCR Rowntree Housing Studies Seminar, University of Bristol, December 1984

Chapter Three

THE DETERIORATION OF PUBLIC SECTOR HOUSING

Ted Cantle*

A crisis in housing is not new. Indeed, it seems it is a cyclical phenomenon, peaking every 20 years or so. And between the peaks housing problems and the resources devoted to them, are seldom subject to a planned and rational approach. We lurch from year to year, from a high point to a low point; responding to some 'new initiative' or another, which amounts to sticking fingers in the dyke as the water level builds up behind it.

There is never any single factor which brings us to another crisis point in housing, but some factors are, of course, more important. After the two World Wars the shortage of dwellings reached very unacceptable levels, and in the 1960s the problem of the slums demanded a rapid and massive response. Now in the 1980s a number of different problems, concerned with the physical condition of the public sector stock, are very evident. Ironically, the larger share of these problems were born out of the responses to earlier crisis - massive public sector housebuilding programme using new designs and techniques were used to meet housing shortage and replacement dwelling programmes in the post-war period. As other contributions to this book make clear, there are many other aspects of the housing problem reaching crisis proportions, and whilst this is largely due to the absolute decline in housing investment, other areas of housing activity have suffered disproportionately because the public sector stock has swallowed an increasing proportion of available resources.

---

* Whilst this chapter draws heavily on the work of the Association of Metropolitan Authorities, the views expressed are not necessarily those of the Association.

The 'Traditional' Local Authority Stock

It has become all too easy to imagine that local authorities' problems are concentrated in, or even limited to, housing built in the post-war period. High rise and deck-access industralised and system built dwellings have received considerable interest from the press and media; and the Government's main response has been to produce the Housing Defects Act 1984, which was directed at the problems of prefabricated reinforced concrete dwellings built prior to 1960. (1) Ronan Point, the block of flats which partly collapsed in 1968, sealing the fate of industrialised house building, has even re-emerged as a potentially dangerous problem. (2) But in March 1985, the Association of Metropolitan Authorities (AMA) published their third report on housing defects, 'Defects in Housing Part 3: Repair and Modernisation of Traditional Built Dwellings', (3) which pointed out that local authorities had been diverted from their ongoing programmes of modernising pre-war dwellings and had been forced to turn their attention to the even more pressing problems in the post-war stock. The AMA estimated that out of some 1,200,000 pre-war dwellings in local authority ownership in England, 450,000 were unmodernised. This, together with major items of repair and remodernisation to other pre-war dwellings, would cost around £8,000 million.

The AMA reported that a survey of their member authorities, which include all but one of the London Boroughs and towns and cities in the major conurbations of England, had revealed that at the current rate of progress it would take 20 years to complete the modernisation of dwellings already around 60 years old.

Furthermore, the rate of progress was slowing down so rapidly that they felt a programme of 50 to 100 years was more likely. Several authorities had abandoned altogether the modernisation of pre-war housing, and more were doing so, or concentrating on the major items of repair only. For some authorities, the prospects were particularly alarming: for example, Birmingham would only modernise 40 dwellings a year out of an unmodernised stock of 24,900 - a programme period of 622 years! Liverpool was unable to improve any of its 10,050 unmodernised dwellings, and Leeds, Manchester and Sheffield had nearly 55,000 dwellings between them left to modernise.

In Scotland there are over 68,000 pre-war dwellings requiring comprehensive modernisation and over

27,000 needing partial modernisation. There are also nearly 200,000 post-war dwellings of traditional construction requiring a comprehensive or partial modernisation; and in both categories over a quarter of a million dwellings require electrical re-wiring alone. For all types of construction, 234,000 dwellings need treatment for condensation and over 80,000 for rising or penetrating damp. (4)

The development of council dwellings in the early 1920s explains some of the present problems. Firstly it must be said that the standard of pre-war purpose built dwellings was very high compared with those built for rack renting prior to 1914. The 1919 Housing Manual (5) put into effect the recommendations of the Tutor Walters Committee of 1918 which was the first serious attempt to define and lay down acceptable housing standards. Recommended space standards were particularly high and every house was to have an internal wc, usually sited off the rear entrance lobby, and facilities for cooking and washing. The dwellings were laid out at a low density, on a varied road pattern, and with large individual gardens.

Unfortunately, the high standards laid down by the 1919 Manual did not last, and were not always put fully into effect when houses were built under the enabling Housing Act (The 'Addison Act') 1919. Space standards, in particular, were reduced by the adoption of the non-parlour house type. This, says Burnett, (6) followed Ministry encouragement and would save around £100 per house - 10% of the total cost. Subsequent Housing Acts specified lower standards, and the immediate post-1919 enthusiasm was quickly and firmly dampened down. Thus, Burnett reports, "The great majority of three bedroomed local authority houses built after 1923 had superficial areas between 750 and 800 sq. ft. instead of 900 sq. ft. proposed in the 1919 Manual". (7)

Bowley (8) has calculated that only 170,000 were constructed for local authorities in England and Wales under the 1919 Act, out of a total of 1,111,700 built prior to 31st March 1939. Many of the 170,000 would be built to below the 1919 standard, but for the remaining 900,000 or so houses built subsequently, standards were almost certainly lower in terms of space.

The reductions in space standards have important repercussions for repair and modernisation programmes, bearing in mind that most of these houses, especially the non-parlour types, have a downstairs wc outside the main rear entrance door and a very small bathroom, usually also downstairs. Furthermore, most houses have only a cramped scullery (now generally used as a kitchen). In the meanest type, the bath was located in the scullery itself, under a wooden cover. The provision of new amenities, including proper kitchen and sanitary fittings, can therefore be difficult and expensive - as the AMA report (9) points out:

> Parlour houses, because they are generally larger, are easier to adapt. Both parlour and non-parlour types suffer from a very cramped scullery (now used as a kitchen) often with a larder which is not efficient in terms of space used. For parlour types, a good solution may be to create a 'living kitchen' from the scullery and living room, retaining the parlour as a separate sitting room. In non-parlour types, this is not possible, and a satisfactory solution may only be found by combining the scullery and bathroom to form a kitchen, and removing the bathroom to upstairs. This depends upon the bedroom arrangements, and it may be necessary to reduce the size of the house from a 3 bedroom, 5 person to a 3 bedroom, 4 person or even a 2 bedroom house. Given that many non-parlour houses are well below 800 sq. ft., it may be desirable to recognise that reducing occupancy is appropriate. However, this will depend on present occupancy and demand.
>
> Bathrooms in pre-war houses were usually very cramped, often with a small bath. The wc was generally situated outside the external door. To provide an adequate bathroom, let alone two wc's (one in the bathroom) is a real challenge. A larger bedroom may be divided, but this is expensive, especially if new window openings need to be created.

The problems of internal arrangement are central to modernisation plans, and the reduction in space standards has led to major constraints on design solutions and greatly increased the costs involved. Apart from the general unacceptability of the arrangement of these facilities today, most of the fittings are now outworn in any case, and many of the service pipes are in lead, with the plumbing generally antiquated, and insufficient heating and hot water systems. Indeed, many of these dwellings still have the original fittings, or the legacy of them, which consisted of a cooking range in the living room, and perhaps a boiler or furnace in the scullery for hot water.

In addition to the problem of standards pertaining in pre-war council dwellings, there is now an increasingly evident problem of repair. Apart from the sanitary and other fittings, the main structural elements are deteriorating, in some cases rapidly. One of the first and major concerns is the condition of the external walls themselves. Generally, good quality brick is used, but most dwellings built prior to 1930 have solid (9") walls. Rain penetration is therefore able to occur, particularly where mortar joints have been eroded. There is also a problem of absence, or failure, or horizontal damp proof courses at ground level, and external rendering, where used, is now requiring attention. But simple repair of solid external walls is generally insufficient in itself. Thermal insulation will be poor, and condensation is a fairly common problem. A more comprehensive upgrading is therefore generally necessary, though far more expensive. Early cavity-walls are usually thermally more efficient, but if built with a brick inner-skin, may also give rise to condensation problems. Furthermore, cavity wall-tie failure is becoming more prevalent in dwellings of this age due to corrosion of the ties. This causes major structural damage and, in some cases, like for example in Sheffield, has led to the demolition of such properties.

Problems of dampness in floors are also evident due to absence of damp proof membrane, and unstable sub-floor aggregate fill has caused floors to 'heave', and even dislodge the foundations to the external walls. Houses around 60 years old also commonly suffer from roof problems. Slates and tiles slip as nails corrode, and will even de-laminate and break up. No roof-felting was provided, neither was there any thermal insulation.

In addition, the external joinery items will almost certainly need attention, if not replacement. Frequent maintenance and painting of such items will in any event be more expensive in the future, but the sliding-sash windows and ledged and braced doors and the like are draughty and old fashioned.

Internally, apart from the sanitary, cooking and heating appliances and fittings, perhaps the electrical system is the most important item for repair and upgrading. Often only one socket was fitted upstairs, and the cables may well be sheathed in lead or rubber, particularly vulnerable to the heavy overloading which is inevitable today given the vast increase in household appliances. Internal finishes, particularly the plaster to walls and ceilings, may also require considerable repairs, and the original chimneys and flues are seldom appropriate for today's heating appliances.

Most of the repair items are not 'defects' in the sense that we have now come to associate them with premature failures. These 'defects' are repair and replacement items which could have been anticipated in dwellings 60 years old. Indeed, the properties have stood the test of time very well on the whole, and apart from the constraints imposed by the lowering of space standards, are relatively easy and inexpensive to remedy. The AMA found (10) that for their member authorities, the range of costs was £10,000 to £15,000 outside London and £15,000 to £20,000 inside London, with an overall average of £12,500 per dwelling. This may reflect, however, a lowering of standards of repair and modernisation, due to restrictions of capital expenditure by the present Government.

The omission from the AMA's report, however, was a detailed consideration of the estate infrastructure. Perhaps the most difficult problem in this regard is that of inter-war flats. Burnett (11) estimated that less than 100,000 flats were built in London and the other main cities. These are generally medium-rise walk-up blocks with an open deck access, and have a very unattractive appearance. But 'environmental' problems are not limited to appearance alone. Security is a major consideration for the flatted estates, for example. Further, the provision made for motor cars in flatted estates, (and in traditional house layouts), is quite inadequate - narrow roads, no off-street parking and few garages is the norm. And the roads, footpaths, street lighting, drains, sewers, and hard landscape areas, which, of course, are also 60 years old, now require considerable re-investment.

The AMA estimated (12) that there were also around 300,000 traditionally built dwellings constructed prior to 1914, now in the ownership of local authorities in England and Wales. Only Some 20,000 were purpose built and these are concentrated in the cities and often in flatted form. They are generally in worse condition, simply because of their age and even less attractive than the inter-war flats. However, the majority of pre-1914 properties are individually acquired dwellings bought for a multiplicity of purposes and reasons - perhaps for improvement from owners reluctant to invest, or transferred from the highway authority having been acquired to facilitate a road improvement scheme. These 'sundry' properties will also figure in local authorities' bids for capital expenditure.

The post-war traditional stock is also now of an age where a number of quite major items require repair or replacement. Modernisation is, however, not generally necessary on the same scale as for pre-war properties, but as these dwellings are now around 30 years old, authorities are faced with the prospect of trying to 'capitalise' repairs out of an ever-reducing capital budget, or see revenue costs on day to day maintenance go on increasing.

The AMA estimates that there are about 625,000 traditionally built dwellings constructed prior to 1960 in the ownership of English local authorities. These post-war dwellings are generally of much higher space standards than the inter-war stock. This was the result of the higher expectations prevalent again immediately after the war with the Dudley Report of 1944 (13) giving rise to the Housing Manual (14) of the same year. Whereas the Manual recommended a minimum of 900 sq. ft. exclusive of stores the pre-war houses were generally 750-800 sq. ft. for a 3 bedroom 5 person house. Apart from the overall increase in space, the recommended designs had proper kitchen and bathroom provision at the outset. Two wc's were even to be provided in early designs.

Space standards were again reduced soon after the war, with enthusiasm tempered by economic considerations. The average floor area of the standard 3 bedroom 5 person council house fell from 1,050 sq. ft. in 1951 to just under 900 sq. ft. in 1960. (15) Space standards were still above pre-war houses, however, and this together with the new design concepts has meant that few dwellings of the era have required major internal alterations.

The design of the main structural elements of the post-war houses are also a considerable improvement on pre-war standards. Walls are generally of the cavity type, and roofs of traditional constuction have needed little more than routine maintenance. External joinery has not fared so well, however, and metal window frames (especially early non-galvanised types) are proving difficult to maintain. Kitchen and sanitary fittings, after 30 years, may also require replacement, but this is generally undertaken piecemeal. Perhaps more important, early post-war electrical installations have proved inadequate for today's household needs and require replacement. Also, the heating systems are both insubstantial and inefficient and the problem of breakdowns is compounded by the difficulty in obtaining spare parts for what are often obsolete appliances. Another major item is the rainwater goods which are generally asbestos cement, or steel, and replacement schemes are necessary.

It is dwellings of this period which are now fast falling into disrepair. For some, only one item now requires a major repair or replacement scheme. However, most have several items which now need attention, and authorities have been able to give little thought to upgrading standards - such as the installation of central heating and improvement to thermal and noise insulation levels.

34 AMA members (16) reported that some 200,000 traditionally built properties in the early post-war period needed repair and upgrading, but many had simply not been able to consider them due to the backlog of other more pressing problems. However, only 9 AMA authorities had any sort of programme for tackling such dwellings. If nonmetropolitan districts have a problem on a similar scale then 400,000 dwellings could require reinvestment, and if the average cost of £5,000 applicable to AMA members was used, the cost could be fairly modestly estimated at £4,000 million excluding Scotland.

This is a cost which must be added to the estimate for the generally more urgent pre-war repair and modernisation needs - which the AMA has estimated at £6,000 million for the 450,000 dwellings requiring modernisation, and a further £2,000 million on capitalised repairs.

The anticipated costs for Scottish housing are similarly alarming. Assuming similar unit costs of modernisation (i.e. £12,500 full modernisation of pre-war houses and £6,000 for partial modernisation; £8,000 post-war, average for comprehensive and partial schemes; and £1,000 for electrical re-wiring - the only capitalised repair identified) and applying them to the numbers in each category referred to earlier, the total cost is approaching £3,000 million. This is made up of £1,012 million for pre-war improvements; £1,600 million for the post-war stock; and £250 million for re-wiring.

The current cost of dealing with the local authority traditional stock in Great Britain can therefore be estimated at £15,000 million.

'Non-Traditional' Dwellings

'Non-traditional' dwellings are a form of permanent accommodation largely designed and built immediately after the Second World War. There had been some earlier dabbling with these new methods of house construction in the 1920's when 52,000 dwellings were built. (17) The post-war use of non-traditional methods for permanent accomodation was also preceded by the construction of 157,146 temporary single-storey prefabs (of which 32,176 were in Scotland). (18) (In 1975, some 15,000 temporary prefabs were still in use in England and Wales, (19) and many remained in use for much longer than the 10 year life originally envisaged). Given the limited nature of this early experimentation with the new techniques it is not hard to understand how the sudden rise of 'non-trads' has led to the present problems. As many as 500,000 'non-traditional' dwellings were completed in the early post-war period (20) - most between 1947 and 1953.

The concept of non-traditional dwellings was explained in a Government Information Paper (21) thus:

> Non-traditional building may use new or the same basic traditional materials in new ways, employing new techniques in fixing and erection which differ, for instance, from the traditional method of laying by hand one brick, or concrete block, on top of another. In the main, new methods have been applied to alternative systems of walling, employing concrete posts and in-filling panels; thin concrete slabs supported on light structural steel framing; preassembled panels of brickwork; stressed-skin resin-bonded plywood panels, asbestos sheeting in various forms; curtain walling and the like. These are usually produced in a factory and transported to the site, requiring only to be placed and secured in position. Cupboard fittings for kitchens, items of joinery, etc., or even complete dwellings (as exemplified in the post-war temporary bungalow) can also be included under this heading.

In practice, the main difference between 'non-traditional' and 'ordinary' (i.e. traditionally built) dwellings was the use of prefabricated components or in-situ concrete for the external walls, although some systems used a much higher degree of prefabrication including roofs, floors and internal walls.

Non-traditional dwellings were a response to the problems of the day: a massive demand for new housing as a consequence of low activity in the war period, higher rates of household formation, and housing damaged or lost by enemy action on a massive scale; and the traditional building industry was devoid of the necessary skills, there was a shortage of traditional building materials, and a generally high cost low output system of production. The Government had to act - Labour was returned in 1945 after promising to build 240,000 homes each year, (22) and amid generally rising expectations. The Government demanded higher output and would harness the "extensive use of new methods of construction which economise in labour, by standardisation and by the use of labour industrial capacity normally outside the industry" and "prefabrication and other non-traditional forms of construction which make a smaller call on building labour than the traditional methods...will be used to the fullest practicable extent in the construction of permanent houses during the emergency period". (23)

The Government attempted to firmly establish the nontraditional methods and set up the Post-War Building Studies on House Construction; an inter-departmental committee appointed by the Minister of Health, the Secretary of State for Scotland, and the Minister of Works. The committee evaluated house types using nontraditional methods, which had been devised by private and, to a much lesser extent, public developers. Their first report was produced in 1944, (24) and second and third reports followed in 1946 and 1947. (25) The approved list of systems included 46 types using pre-cast and in-situ concrete, 35 sheet-framed houses, and 20 timber framed and solid timber houses. (26) The Government thereby tried to reduce local authority suspicion of the new systems by giving assurances about quality and durability. However, this proved insufficient in itself and central government were prepared to give local authorities inducements to use nontraditional types and to set targets:

> The Government established the market for permanent 'non-trads' by compelling local authorities to take these houses to the extent of a new certain percentage of their annual housing programmes. The new Towns were obliged also to provide 15% of their annual programmes as non-traditional houses. (27)

The Conservative Government, which took office in 1951, was equally enthusiastic about the new methods, and indeed this was necessarily so in view of its more ambitious housebuilding targets of 300,000 per annum. (28) The Ministry Circular issued in 1952 (29) explained:

> There will be delay in relieving the hardship and anxiety of those on waiting lists and others in need of homes unless the maximum output of traditional houses attainable in present circumstances is supplemented by the use of established nontraditional methods.

The same Circular offered higher programme instalments and additional allocations for authorities using the new methods.

The decline of nontraditional methods began in 1954 after the Conservative Government had met its house building target, pressure began to ease on the housing front, and the traditional building industry was recovering. When Government sponsorship and inducements to builders and local authorities were terminated, nontraditional methods began to fade:

> As the availability of traditional labour and materials improved - from about 1954 on - the Government gradually relaxed its efforts on behalf of the 'nontrads' and only those that could compete directly in the open market with traditional building materials have survived." (30)

By this time, however, 116,601 'nontrads' were included in approved tenders in Scotland (1st January 1945 to 31st December 1954) in 28 name types; (31) and between 1st April 1945 and 31st December 1955, 339,680 dwellings were in tenders approved in England and Wales. (32) The latter included 18 named types, with a further 14 types found by the AMA which were not identified at the time. (33)

Considering the history of nontraditional dwellings, it is not surprising that so many have now failed, and require considerable reinvestment. The AMA has estimated (34) that the cost of rectification will be at least £5,000 million for the 500,000 'nontrads' that it believes were eventually built throughout Great Britain (the specific statistical returns were terminated in 1954 for Scotland, and 1955 for England and Wales). The central government sponsorship of nontraditional methods has also led to claims of compensation by the local authorities now faced with the massive costs associated with repair, and there can be little doubt that central government efforts at appraisal were inadequate and superficial. The same mistake - rapid growth in house building targets, and the use of new and untried materials and techniques, and superficial appraisal by central government agencies - were repeated in the 1960s leading to probably even greater problems. And there is a danger that they will be repeated again: firstly, when house building in the public sector increases - as it inevitably must; and secondly, in the remedial systems being developed to repair 'nontraditional' and industrialised dwelling types. The 'repair systems' are themselves untried and untested, and there is a flurry of activity from producers and developers to establish their particular brand in the market which is desperate for 'solutions'.

The defects associated with 'nontraditional' dwellings are many and various, but it is also the case that a number of items require modernisation, simply on account of 'fair wear and tear' over 30 years of life. Indeed, these items are similar to those discussed above in relation to the post-war traditional stock - electrical rewiring, plumbing, and sanitary fittings, heating and kitchen appliances, window frames, rainwater goods, etc. 'Nontraditional' dwellings, however, made far greater use of lightweight materials (a prerequisite of prefabrication) and generally proved less durable. For example, roofs were often covered in metal or asbestos cement sheets, first floor external walls (and perhaps the ground floor as well) may have also been clad in sheet materials or panels, and framed structures were often clad, or lined, internally rather than plastered.

It is the main structural elements that have warranted special attention, and the deterioration of these elements has meant that some dwellings have had to be demolished, and many are generally unmortgageable. The full extent of the problems have not yet been fully recognised, however. The AMA report 'Defects in Housing Part 1: Nontraditional Dwellings of the 1940s and 1950s (35) called into question _all_ types of nontraditional dwellings and specifically drew attention to structural problems of an advanced state in certain types of steel frame and timber frame houses, as well as the prefabricated reinforced concrete (PRC) types. The Government, however has consistently refused to recognise the full extent of the problems with 'nontraditional' dwellings, and has only reluctantly taken the measures to date, constantly playing down and avoiding the full financial implications.

Government action has been almost entirely confined to PRC types. This began with a suggestion from the DOE that defects could arise in the 26,000 Airey dwellings in May 1981. (36) This was followed sixteen months later by repair grants, or repurchase by local authorities, and in the same month (September 1982) a warning from the DOE about the possible problems in the 15,000 Orlit dwellings. Gradually, more and more types of PRC dwellings were added to the list of suspect PRC types under investigation by the DOE's Building Research Establishment (BRE). This culminated in a national scheme of assistance to the private owners of PRC dwellings, (along the same lines as the Airey house voluntary scheme), established by the Housing Defects Act 1984. This Act allows any type of defective dwelling to be designated which has an inherent 'qualifying defect', which is generally known about, resulting in valuation problems. However, Ministers have attempted to gear the Act to the specific problems of PRC houses, other types have not been designated, and the DOE guidance circular on the Act (37) seems to exclude even the possibility of designation of other types. This narrow view is understandable in the context of the Government's continued and relentless assault on housing capital expenditure. To recognise the full extent of the problems would obviously bring into question their expenditure plans. Nevertheless, even the problems which they have recognised in the PRC dwellings are very substantial. In the explanatory memorandum to the Housing Defects Bill (38) the Government estimated that the cost of dealing with the 16,500 PRC dwellings in private ownership by repair or repurchase would be £175-£250 million. However, the cost of dealing with the further 153,500 (39) PRC dwellings in public ownership, using the same repair cost per unit of £14,000 would be £2,150 million - and this does not include other items of repair and modernisation, which are not part of the 'qualifying defect', but are now needing attention. The total cost of dealing with these problems is likely to be at least £20,000 per unit, or over £3,000 million, for all PRC dwellings in public ownership.

The types of PRC dwelling designated under the Housing Defects Act on the 1st December 1984 or shortly afterwards, (40) are as follows:-

Airey
Boot
Cornish Unit
Dorran
Dyke
Gregory
Hamish Cross
Myton
Newland
Orlit
Parkinson Frame
Reema Hollow Panel
Schindler and Hawksley SGS
Stent
Stonecrete
Stour
Tarran
Underdown
Unity and Butterley
Waller
Wates
Wessex
Winget
Woolaway

The 'qualifying defect' refers to the problem of the rusting of the metal reinforcement embedded in the concrete posts and panels. (41) As the metal rusts, it expands and cracks the concrete, and in advanced stages, threatens the structural integrity of the dwelling. Some repairs systems rely upon the neutralising of the rusting process and providing a more impermeable layer of concrete cover to the reinforcement. A better, but more expensive repair, is to replace the load bearing PRC elements with traditional brick and blockwork.

The PRC types also suffer from a number of other problems: corrosion of metal wall and panel tiles; inadequate weather seals at panel junctions; condensation of both interstitial nature and surface condensation, because of the poor thermal insulation qualities; inadequate fire stopping between ground and first floors and between separate dwellings; and a number of more minor but potentially serious problems.

Other types of concrete dwellings such as Smith (42) and Mundic Block (43) are also evident, and no-fines concrete dwellings which were particularly prevalent in this period suffer from condensation and other problems, (44) and in common with PRC types, the render finish is particularly vulnerable, 'shelling-off' almost completely in some cases, allowing rain to penetrate.

The steel and timber frame varieties of 'non-traditional' dwellings do have evident faults and are being repaired by local authorities as resources permit. The deterioration of frames has begun in some types, and external claddings - steel, timber and asbestos cement - now generally require attention. The lack of cavity fire-stopping has been fairly widely reported, particularly in the BISF type, and poor insulation and compaction of insulation quilts have contributed to cold dwellings which are difficult to heat. Other structural problems include: lack of cavity ties between frame and brickwork or cladding outer skin; lack of fire-stopping between dwellings; and unsafe gable ends.

The cost of rectification was, as stated above, estimated by the AMA at £5,000 million, but these figures were provided before all of the faults were identified and investigated by authorities. In the light of the costs now pertaining, especially in respect of the PRC types, this may turn out to be something of an underestimation.

Industrialised and System Built Dwellings of the 1960s and 1970s

The symbol of defective local authority housing has been the 1960s high rise block, and to a lesser extent, deck access dwellings of the same period. There are, of course, very substantial problems here but as we have noted, it is very far from being the case that all problems are concentrated in these types. Reactions to defects and physical problems with these dwellings has been conditioned by the fact that the form of development itself has gradually been far less acceptable, particularly where there is a concentration of family accommodation in this form. This is very much in contrast with the traditionally built dwellings, and the post-war nontraditional types, even though the latter can sometimes look rather unusual and flimsy. Up until the 1960s explosion of high rise, and other high density dwellings, most dwellings were constructed in traditionally laid out estates with individual private access and gardens. The 1960s produced new built forms, revolutionalised housing layouts and environments, and created new skylines and street scenes. Consideration of the <u>physical</u> defects is therefore difficult to disentangle from the emotional and social reactions to the built form. Certainly, it is impossible to consider solutions to the physical problems without addressing the more intractable difficulties of social acceptability of the estates as a whole.

It is also the case that the majority of industrialised dwellings were built in a low rise form, and the problems that exist in these types have tended to be overlooked, or overshadowed by their higher-rise contemporaries. Accurate statistics are, however, difficult to come by: the Ministry of Housing and Local Government (and later the Department of the Environment) kept statistics for industrialised house building only between 1960 and 1979, and even the comprehensive survey which the DOE carried out in May 1985 is unlikely to have revealed the full extent of the number and nature of problems, due to the difficulty authorities have in identification. Furthermore, the industrialised building statistics that were kept are, according to the AMA, somewhat unreliable and they identify several systems where building activity has been underrecorded. In their report 'Defects in Housing Part 2: Industrialised and System Built Dwellings of the 1960s and 1970s', (45) the AMA go on to estimate that whilst only around 600,000 dwellings could be specifically identified in statistical data of this period, for Great Britain as a whole, it was likely that the number actually built would be in the region of one million units of accommodation.

A more recent survey has shed some light on high rise housing. 'Tower Blocks' by R Anderson et al (46) is the result of a survey of all housing authorities in Great Britain for which a response was received from 435 (94%). This found that there were 4,570 high rise blocks owned by authorities (high rise is defined as six storeys or more), and these blocks contained over 300,000 dwelling units. High rise blocks tend to be concentrated, according to the survey, in metropolitan areas, particularly Greater London. Ten authorities have more than 100 tower blocks, one in excess of 400. Estimates of the number of deck access units are less reliable with figures of 75,000 to 150,000 being suggested. (47)

The development of industrialised dwellings was, like the nontraditional programme of the 1940s and 1950s, a response to a housing problem again reaching crisis proportions; and central governments - both Conservative and Labour - saw it as a necessary means of rapidly boosting output over and above the capacity of traditional building industry. Industrialised building has been defined thus:

> The term 'industrialisation' covers all measures needed to enable the industry to work more like a factory industry. For the industry, this means not only new materials and construction techniques, the use of dry processes, increased mechanisation of site processes, and the manufacture of large components under factory conditions of production and quality control; but also improved management techniques, the correlation of design and production, improved control of the selection and delivery of materials, and better organisation of operations on site. Not least, industrialised building entails training teams to work in an organised fashion on long runs of repetitive work, whether the men are using new skills or old. For this purpose, industrialised building can include schemes using fully rationalised traditional methods. (48)

System building may use traditional methods, but usually relies heavily on industrialised techniques - it provides a design and working method (often with specially made plant and components) - and may be proprietary or nonproprietary, although the majority were linked with a particular builder or developer.

Central government therefore felt industrialised building to be especially suitable for mass production, along factory lines, not limited by availability of skilled labour or traditional building materials, and less disrupted by inclement weather. It must be recognised that the housing problems facing the nation were again exceptionally severe. Unfortunately, the progress in the early 1950s, in terms of public sector housebuilding, had not been sustained and by 1960 was half its peak (1953) level of 229,000 completions in Great Britain. (49) The earlier crisis was essentially the legacy of war damage and having had virtually no housing activity over a six year period, together with higher rates of household formation. The new crisis was primarily about the slums. The Government White Paper of 1961 (50) noted that there had been 847,000 slums in 1955 (compared with annual clearance of only 30,000), but it was soon appreciated that this was a considerable under-estimate. Indeed, by 1965, another White Paper (51) said that needs existing were:

(i) 1 million (dwellings) to replace unfit

houses already identified as slums;
ii) up to 2 million more to replace old houses not yet slums, but not worth improving.

and further problems were identified too:

iii) about 70,000 to overcome shortages...and for mobility;
iv) 30,000 a year to replace the loss caused by demolition; and
v) 150,000 a year to keep up with new households being formed in the rising population.

This view is consistent with the Labour Government's election promise to build 500,000 houses a year - a target which may have contributed to the downfall of the Conservative Government in 1964, who had pledged 350,000 dwellings per annum.

Both Conservative and Labour Governments were committed to high levels of house building and industrialisation. They ensured local authorities complied, through administrative and financial measures. The Housing Subsidies Act 1956 provided for higher subsidies for higher buildings and this continued in the Housing Act 1961. Storey height was linked to higher densities, and density was an important factor, given that the slum houses that were to be replaced were themselves built at high density. The contribution which could be made by overspill estates and new towns, relocating people from the city centres, was seen as being modest from the outset. This was partly because some cities were opposed to such relocation, and particularly removing people from areas close to their employment; and because of a more general concern about the need to limit urban sprawl and protect the Green Belt. There were therefore particular pressures to build high and dense; but there was a more pervasive influence upon local authorities to adopt industrialised techniques.

The AMA has explained (52) how this influence was used: the Government expected local authorities to have 25% of output by industrialised methods by 1963, and the target of the next (Labour) Government was to increase industrialisation to 40% by 1970; the Ministry of Housing and Local Government (MHLG) also gave practical assistance to authorities, grouping contracts together, negotiating prices with contractors, speeding up approvals, and by giving additional allocations to authorities which co-operated with the industrialised building 'drive'; Goverment departments were also working with developers to bring forward schemes and the MHLG even produced their own system, known as 5M. Technical innovations, such as the agreement of preferred dimensions to enable the interchange of components and the introduction of a national scheme of by-laws were designed to smooth the path of progress. Perhaps most significantly, in the context of present discussions over the cost of repairs and responsibility for them, the Government established the National Building Agency (NBA) in 1964. It was specifically charged with promoting industrialised building and to appraise new methods. (53) Systems were then subject to NBA approval and 89 appraisal certificates were given to warrant that the design "is sound and suitable for a 60 year loan sanction". (54) The NBA appraisals have since been shown to have been superficial and highly misleading, and had it not been wound up, the Agency may well have found itself subject to litigation by authorities in receipt of appraisal certificates for dwellings which are now very defective, and in some cases, facing demolition.

The AMA have estimated that the cost of rectification of an estimated one million dwellings in this category will be between £3,750 and £5,000 million, and believe that 10,000 dwellings, on average only 15 years old, have been demolished at a loss of £300 million. (55) In the light of more recent developments, however, it is likely that both the final cost and the numbers demolished, or scheduled for demolition, will rise. Revenue costs are also very considerable - like the older traditional and post-war non-traditional stock, the <u>new</u> debts associated with repair and modernisation must be serviced, generally over 20 years. However, dwellings which are only 15 years old have considerable outstanding <u>old</u> debt charges, to be repaid over the next 45 years without any rent income to offset these charges. For example, Leeds demolished 1,249 YDG dwellings in 1984. The debt outstanding was £4.76 million, and demolition and compensation costs added a further £650,000. Leeds are now faced with annual debt charges of around £800,000 each year for the next 45 years.

The physical defects associated with industrialised dwellings are many and varied. Most worrying are those defects which threaten the structural integrity of blocks of flats, although they may well be visible internally or externally, and may have little effect upon the tenant's enjoyment of the property. Ronan Point, a symbol of the failure of system building when the partial collapse occurred in 1968, has again highlighted new structural problems. Ronan

Point is among over 8,000 dwelling units in the Taylor Woodrow Anglian system. Problems in the jointing of floors and walls have now again been found, and the DOE fear that some of the problems may be inherent in the system. Local authorities have, accordingly, been advised of these problems. (56) The bowing outwards of external panels is sometimes visible internally, as gaps appear vertically between flats. The structural connections, however, are altogether more difficult to check. Similarly, dwellings built in the Bison Wall Frame system have been found (57) with inadequate or ineffective connections between walls and floor panels. In addition, the ties between inner and outer leaves are lacking altogether, insufficient, or are of unsuitable material. Similar problems have also been found in other systems, but a more general problem is the inadequate concrete cover to metal reinforcement or the presence of chlorides in concrete which brings about the rusting of the reinforcement. As metal rusts, so it expands, and breaks upon the concrete, causing it to 'spall' - i.e. break away altogether. Lumps of concrete spalling off tower blocks are, of course, very dangerous and authorities have had to erect protective canopies as an interim safety measure.

The systems have had a general problem of watertightness, particularly in high rise construction. Panel systems have been vulnerable at the joints, and movement between panels, and elements of dissimilar materials has exacerbated jointing problems. But the exposure conditions in high rise flats have tested even the better built systems, and it has been found that even the slightest hairline crack, say around window frames, can allow moisture to penetrate. Moisture can also penetrate through parapets, lift housings, and even through well constructed brickwork panels. Rain penetration is also prevalent with flat roofs, and low-pitched roofs, which are fairly common in low rise industrialised units where prefabrication and lightness of construction was a major factor.

Condensation may be associated with structural problems, such as poor insulation and 'cold-bridging' (and the colding effect of weather penetration is also a factor), but also because of the totally inadequate forms of heating (e.g. electric ceiling heating), which were cheap to install and did not require a flue, but are very expensive to run.

Differential and thermal movement is also widely experienced. The lack of expansion joints between dissimilar materials, can cause the structure to crack and crumble - a common problem is that of concrete framed buildings with brick infill cladding panels - and staircases, link bridges, balconies and so on can be seriously affected. Certain products and materials, such as metal flues, had not been tried and tested and have not proved durable and chemical reaction problems were subsequently found with high alumina cement and other additives to concrete. The use of asbestos, which is particularly widely used in lightweight systems, has created further problems and required encapsulation or removal programmes, for both inside dwellings and within the estate environment - e.g. on deck access walkways. Many other problems - lack of sound insulation, poorly designed district heating schemes, premature rotting of external joinery and so on - have added to the catalogue of disasters, and added to the general resident dissatisfaction with the form of development which is also evident in many cases.

The built form of industrialised dwellings is a further factor of considerable importance. In contrast to nontraditional dwellings of the 1940s and 1950s and the earlier traditional stock, it is far less likely to be in semidetached form and laid out on a low density estate with a traditional road pattern. The national survey of high-rise local authority housing published in 1985, referred to above, (58) revealed that there are over 300,000 dwellings in high-rise form. Further, over 70,000 deck access units in the UK have been identified by research carried out between 1978 and 1982. (59) High and medium rise forms were developed partly because of the higher densities achieved, and also because of the building advantages - repetition of floor plans and components and so on - and because of the emergence of a new architectural movement. But the design experimentation, and even density considerations, also extended to much of the low rise industrialised housing. Long terraces, often at right angles to the road, and with open-plan frontages and grouped parking and garaging - which may be some distance from the dwelling - has in itself proved an unpopular form.

Physical defects and an unpopular estate environment has often combined to make many of the industrialised dwellings the last choice of would-be council house tenants. Tenant dissatisfaction, high levels of turnover and vacancies, vandalism - particularly in the estate 'no man's lands', and the concentration of families which find such conditions the most difficult to cope with, combine to produce a spiral of despair and deterioration.

A Programme of Renewal?

The development of a programme to tackle the problems now emerging in the local authority housing stock has become vital and urgent. Demolition and rebuilding may well be used more frequently in the future as a remedy, given both the cost of repairs and the social unacceptability of dwelling types even after repairs have been completed. But a massive repairs programme is particularly urgent if further deterioration in the structure of such dwellings is to be avoided. Despite Government prevarication and continual denials by Ministers that the problem has reached crisis proportions, a consensus is beginning to emerge and an acceptance that billions of pounds will be required within a fairly short timescale.

The English House Condition Survey (EHCS) (60) has shown itself to be an inadequate tool in measuring the extent of local authority problems, but in spite of the small sample size, the weighting to older private sector dwellings, and the preoccupation with lack of amenities and disrepair in traditional dwellings, the 1981 results were revealing. Local authorities in England had an estimated 67,000 'unfit' dwellings; 143,000 lacking basic amenities; and 608,000 dwellings with repair costs over £2,500 (1981 prices). The three reports on defects in housing published by the AMA (61) suggest much higher figures. The AMA suggests that there are up to two million dwellings of traditional construction, 'non-traditional' types and industrialised and system built houses which are defective to some degree and will cost £19 billion to remedy. The higher figures are to some extent supported by individual authorities in England. The aggregate HIP 1 figures for 1984 (62) show that there are 78,000 'unfit' dwellings, 112,000 lacking basic amenities and 853,000 'nonstandard' properties in need of repairs in excess of £3,000 outside London, and £4,200 in Greater London. However, as the 'nonstandard' figure for the same authorities was only 710,000 in the previous year, it suggests that authorities, have not yet themselves discovered the full extent of the problem. A comprehensive survey carried out by the DOE in May 1985, in response to intensive pressure from local authorities can be expected to confirm the enormity of the problem, as more and more authorities complete survey work. For example, Manchester, Leeds and Sheffield say their stock of local authority dwellings requires expenditure of £400 or £500 million in each case and several London boroughs have given a similar or even higher figure. Glasgow and Birmingham have far more extensive problems however. Despite the work undertaken to date, by various agencies, it is likely to be some years before the full extent of the problem is known, and the defects have been identified and properly investigated and costed.

The problem of premature deterioration of the housing stock is not confined to Great Britain. A recent conference in the Netherlands (63) highlighted growing concern in a number of Continental countries. Britain, in fact, used and adapted systems, particularly from France and Scandinavia, in the system building era of the 1960s and 1970s. Furthermore, there is no line drawn around housing activity, and problems are beginning to appear in commercial and industrial and other buildings. New Scotland Yard, for example, has had a protective canopy around it for some time to catch any cladding which may slip. Ironically, the Department of the Environment's own headquarters in Marsham Street, London, is also showing signs of carbonation and repairs will be necessary. (64) Bickerdike Allen, an architectural practice specialising in remedial works suggested that the cost of rectifying commercial buildings would be £1,000 million. (65) In other words, it is a 'building problem' rather than a 'housing problem', and will only be completely solved by the development of better techniques, improved design and supervision, and proper assessment of products. Public funded research and development work in the construction industry is worth only around £13 million per annum out of a total yearly expenditure of £22 billion (66) (at 1980 constant prices). The paucity of such expenditure is at the very root of our present problems with the post-war stock. Indeed, local authorities, and private owners, are using remedial systems, which rely upon 'nontraditional' techniques, and have been inadequately tried and tested, and with little guarantee of long term durability.

The AMA quantifies the problem of the local authority stock at £19 billion. This figure relates to both industrialised and 'nontraditional' dwellings of post-war period, and largely to the prewar traditional stock. It is also largely confined to English and Welsh authorities. If the costs of tackling the prewar traditional stock in Scotland are added, and some of the more pressing postwar problems arising in the traditional stock, the cost rises to £25 billion. Further, the AMA themselves believe that the problem of industrialised stock may well have been underestimated, and more problems will yet be found.

However, the problem of the stock condition is not static. The Audit Commission estimates that the backlog of disrepair is growing at around £900 million per annum. (67) The Commission's estimate is based on the modest guidelines of the Royal Institution of Chartered Surveyors who suggest that routine maintenance should amount to 1.8% of the insured value of the property. This implies annual maintenance expenditure of £450 per dwelling on average, and compares with the likely figure of £280 per dwelling in 1984/85 - a shortfall of £170 before any efficiency considerations are taken into account. Bearing in mind the unusual design and construction techniques employed on many local authority dwellings, this will be a considerable underestimate.

Despite the Government's enormous reductions in housing capital expenditure, described in Chapter 2, local authorities have devoted a larger share of their capital programmes to renovating their own stock. In 1983/84, 36% of local authority capital investment in England was on renovation of their stock compared with 17% (of a much larger programme in real terms) in 1977/78; but this has become increasingly out of step with the allocation system, as only 8% of the General Needs Indicators related to local authority stock. (68)

The lack of resources is at the root of the problem. Local authorities are quite unable to rectify defects in the postwar stock; keep a modernisation programme moving at an acceptable rate for the traditional older stock; and keep stock in good repair and prevent continuing deterioration. Such work as is being done is at the expense of other housing activities such as new building for the homeless and badly housed, and providing improvement grant assistance and clearance schemes for the worst of the private older housing stock.

The Government's response to the problems have been firstly, to reinforce and commend the work of the Priority Estates Project and emphasise the management problems, rather than the physical problems which more obviously require an injection of new financial resources. The 'Priority Estate' approach is by definition selective, and appears only to entail the concentration of existing financial and staff

resources, together with assistance from other agencies such as the police, social and community service workers, and
directed, usually from within existing budgets, and occasionally from other programmes, such as the Urban Development Grant. Many of the lessons learnt in management techniques, tenant involvement, and so on, have been worthwhile, but resources are needed on a much larger and widespread basis; and the principal lesson has undoubtedly been quite simply that 'investment pays off'. Estate decline can be arrested, the quality of life improved, the number of vacant dwellings reduced, and vandalism even eliminated. Furthermore, the physical, social and environmental aspects of the estate cannot be neatly divided up - they can combine to produce an undesirable, unpopular, and unattractive place to live.

The Government's second response has been to extol the virtues of 'private finance'. This term is almost meaningless as most local authority capital expenditure on housing is obtained from private sources. The use of 'private finance' will necessitate the privatising of the dwellings concerned, irrespective of the source of the money. (69) This situation is brought about by the Government's irrational approach to housing finance, but sits quite happily with their doctrinaire housing policy in general - the main objective of which appears to be that owner occupation is a desirable in itself irrespective of the benefits to the condition of the housing stock. The number of dwellings sold by English local authorities (excluding sales to sitting tenants) however is marginal - under 20,000 in the three year period 1981/82 to 1983/84. (70) This is hardly surprising given that most authorities are under such pressure to provide additional accommodation to meet expressed needs in their area, that save for the most exceptional cases the sale of whole estates is impossible to even consider given that vacant possession will be demanded for most of the dwellings concerned.

As the better quality stock continues to be sold to sitting tenants, albeit at a much reduced rate, and the deterioration continues, estates become even less attractive, disaffection grows, and the problem is compounded. There are, of course, many technical, social and operational difficulties to overcome, and there will never be one solution to meet all of the problems. However, a massive injection of capital is clearly essential and very urgent. The money must be spent wisely and imaginatively, and with a new sensitivity - but it must _be_ spent. As the problems mount, there is a real danger that this Government, or a new Government (if their obduracy continues), may be panicked by conditions that can simply no longer be ignored, or even by a tragedy or disaster. We have not yet learnt the lessons of our postwar housing history, that dealing with a housing crisis by the wrong means - even if it seems the only one available - may serve only to sow the seeds of another housing crisis in the years to come. Given that the justified sustained growth in housing activity has been denied in recent years, it is hard to see how such a scenario can now be avoided.

NOTES AND REFERENCES

(1) Housing Defects Act 1984: Assistance for Eligible Private Owners of Prefabricated Reinforced Concrete Homes Designed Before 1960, DOE circular 28/84, HMSO, 7 November 1984.

(2) 'Ronan Point: Summary of BRE Findings Announced', DOE Press Release, DOE, London, 19 February 1983.

(2) Defects in Housing Part 3: Repair and Modernisation of Traditional Built Dwellings, AMA, London, 1983.

(4) Parliamentary Written Answer, Hansard, 19 April, 1985, col. 283.

(5) Manual of State-Aided Schemes, Local Government Board, 1919.

(6) John Burnett, A Social History of Housing 1815-1970, Methuen, London, 1978. p. 223.

(7) Ibid., p. 227.

(8) Marian Bowley, Housing and the State 1919-1944, George Allen and Unwin, London, p. 271.

(9) Defects in Housing Part 3, AMA op. cit. p. 18.

(10) Ibid., p. 27.

(11) John Burnett, op. cit. p. 231.

(12) Defects in Housing Part 3, AMA op. cit. p. 4.

(13) The Design of Dwellings, Report of a Ministry of Health Central Housing Advisory Committee, chaired by the Earl of Dudley, 1944.

(14) Housing Manual, Ministry of Health, 1944.

(15) D V Donnison, The Government of Housing, Penguin Books, 1967, p. 167.

(16) Defects in Housing Part 3, AMA op. cit. p. 29.

(17) Defects in Housing Part 1: Non-Traditional Dwellings of the 1940s and 1950s, AMA, London, 1983,

p. 4.
(18) Houses: The Next Step. Government White Paper, Cmnd 8996, HMSO, 1953.
(19) Parliamentary Written Answer, Hansard, 21 April 1983.
(20) Defects in Housing Part 1, AMA op. cit. p. 13.
(21) Housing in Britain, Central Office of Information, Reference Pamphlet no. 41, HMSO, 1960.
(22) John Burnett, op. cit. p. 277.
(23) Housing, Government White Paper, Cmd 6609, HMSO, 1945.
(24) Post War Building Studies No. 1: Housing Construction. Report by an Interdepartmental Committee Appointed by the Minister of Health, The Secretary of State for Scotland, and the Minister of Works, London, HMSO, 1944.
(25) Post War Building Studies No. 23: House Construction - Second Report Report by an Interdepartmental Committee appointed by the Minister of Health, the Secretary of State for Scotland, and the Minister of Works, London, HMSO, 1946. Postwar Building Studies No. 25: House Construction - Third Report Report by an Interdepartmental Committee appointed by the Minister of Health, the Secretary of State for Scotland and the Minister of Works, London, HMSO, 1947.
(26) Defects in Housing Part 1: Nontraditional Dwellings of the 1940s and 1950s and 1960s - Technical Appendices, AMA, London, 1983.
(27) A W Cleeve Barr, Public Authority Housing, Batsford, 1958.
(28) John R Short, Housing in Britain: The Postwar Experience, Methuen, London, 1982, p. 47.
(29) Expansion of Housing Programme, Ministry of Housing and Local Government (MHLG), Circular no. 28/52 HMSO, 1952.
(30) A W Cleeve Barr, op. cit.
(31) Housing Return for Scotland 1954, Cmd 9366, HMSO, 1955.
(32) Housing Return for England and Wales 1955, Cmd 9681, HMSO, 1956.
(33) Defects in Housing Part 1, AMA op. cit. p. 13.
(34) Ibid., p. 26.
(35) Ibid.
(36) Letter to Local Authorities from the Department of the Environment, 14 May 1981.
(37) DOE Circular 28/84, op. cit.
(38) Housing Defects Bill, published 5 April 1984.

(39) *Parliamentary Written Answer*, Hansard, 8 May 1985, vol. 78 no. 112.

(40) DOE Circular 28/84, op. cit.

(41) Ibid., p. 2.

(42) *Parliamentary Debate* on the Housing Defects Bill, Hansard, 26th July 1984. col. 1299.

(43) Parliamentary Written Answer, Hansard, 4 May 1983.

(44) *See Defects in Housing Part 1* op. cit. for a fuller discussion of defects in PRC types of dwelling.

(45) *Defects in Housing Part 2: Industrialised and System Built Dwellings of the 1960s and 1970s*, AMA, London, 1984, pp. 24-26.

(46) R Anderson, M A Bulos and S R Walker *Tower Blocks*, Institute of Housing/Polytechnic of the South Bank, London, 1984, p.6.

(47) Chris Bacon, 'Deck Access Housing', *Housing and Planning Review*, vol. 40, no. 2, April 1985, pp. 18-21. Also see 'Deck Access Disaster', papers for a conference in Hulme, Manchester, Manchester City Council, 22 February 1985.

(48) *Housing and Construction Statistics: Notes and Definition Supplement*, HMSO, 1979.

(49) S Merrett, *State Housing in Britain*, Routledge, Keagan and Paul, 1979

(50) *Housing in England and Wales*, Government White Paper, Cmnd 1290, HMSO, 1961.

(51) *Housing Programme 1965-70*, Government White Paper, Cmnd 2838, HMSO, 1965.

(52) *Defects in Housing Part 2*, AMA op. cit. p. 11.

(53) Ibid., p. 15.

(54) Ibid., p. 21.

(55) Ibid., p. 47.

(56) *Large Panel Systems: The Structure of Ronan Point and Other Taylor Woodrow Anglian Buildings*, Building Research Establishement, DOE, 1985.

(57) Letter of Local Authorities from the Department of the Environment, Bison Wall Frame Flats and Houses, 5 October 1983.

(58) Anderson et. al., op. cit. p.6.

(59) Chris Bacon, op. cit.

(60) *English House Condition Survey: Part 1 Report of the Physical Condition Survey*, Housing Survey Report no. 12, DOE, London, HMSO, 1982.

(61) *Defects in Housing Part 1, 2 and 3*, op. cit.

(62) Aggregation of Housing Needs Appraisal Forms (HIPI), for English housing authorities, DOE, 1984.

(63) *Postwar Housing in Trouble*, International

Federation of Housing and Planning, papers presented at the Congress in Delft 4-5 October, 1984, Delft University Press, 1985.

(64) Parliamentary Written Answer, Hansard, 20 March 1985.

(65) Press Statement issued on behalf of Wigglesworth (Insurance) Ltd, London SE1, August 1984.

(66) John Kotz, Research and Development in the Construction Industry, Paper to the Public Works Congress, Birmingham, 28 November 1984.

(67) Capital Expenditure Controls in Local Government in England, The Audit Commission for Local Authorities in England and Wales, HMSO, London, 1985, p. 26

(68) Ted Cantle, 'The Renewal of Council Housing', The Planner - (Journal of the Royal Town Planning Institute) vol. 70. no. 4 April 1984, pp. 16-18.

(69) Ted Cantle, 'Housing: The Myth of Private Finance, Municipal Journal, 24 May 1985.

(70) Parliamentary Written Answer, Hansard, 3 April 1985, col. 656.

Chapter Four

HOUSING RENEWAL: PRIVATISATION AND BEYOND

Mike Gibson

A long delayed Green Paper on private housing improvement was published in May 1985. After six years of tinkering with the grant-based subsidy system, these proposals embody the full application of the Thatcher government's housing philosophy to this area of policy. Taken together with the changing role of housing associations and developments in the renewal of council housing the Green Paper clearly signposts the way to the privatisation of housing renewal.

In the early 1980s, against a complex background of stop-go management of public sector investment in housing, civil unrest and mounting evidence of urban decay, the government engineered a brief, but spectacular, boom in grant-aided renovation and instituted a review of long term policy. Paralleled by the collapse of council house building and slum clearance programmes, this boom was essentially the product of political opportunism and expendiency. However, it restructured the overall housing renewal programme in favour of investment in private housing. The second Thatcher administration is now getting to grips with the legacy of opportunism and expediency. The housing policy developments embodied in the Green Paper in the context of the impending restoration of the Treasury's control over local authority housing investment are intended to reduce public sector support for the renewal of private owner housing to the lowest politically acceptable level. Related developments in the housing association sector do not bode well for the inner cities. The crisis of underinvestment in council housing is deepening as attempts to substitute private finance for the withdrawal of public funds generate interesting innovations, but inadequate resources in relation to the burgeoning scale of known needs.

The starting point of this essay is an analysis

of trends in housing stock condition and the changing function of an expanding core of bad housing. The evolution of Conservative housing renewal policy since 1979 is then charted and its impact assessed. This establishes the context for a critical discussion of the Green Paper proposals. The proposals, alongside developments in the housing association and council housing sectors, are bringing housing renewal policies to a crossroads and it is argued that alternatives to the privatisation scenario need to be developed with some urgency.

AN EXPANDING CORE OF BAD HOUSING

Trends in housing stock condition and processes or urban decay have become appreciably clearer with the publication of the results of the 1981 English House Condition Survey (EHCS), (1) supplemented by more recent surveys of defects in council housing published by the Association of Metropolitan Authorities (AMA), (2) and the results of local 'booster' surveys undertaken by an increasing number of housing authorities. (3) Whilst this improving data base still has important limitations, it has been sufficiently revealing to prompt shifts in perceptions of housing problems. But there are competing interpretations of 'the facts'.

The Rising Tide of Disrepair

The changes in the backlog of poor housing conditions identified in the EHCS are illustrated in Figure 1. Superficially these figures can be used to support the view that the decade 1971-1981 was a significant improvement and that although progress slowed in the late 1970s, it was still discernible. Hence in the parliamentary debate on the cuts in improvement grants made soon after the 1983 election, Ian Gow, Minister for Housing and Construction asserted that:

> the truth is that in recent years we have seen a fundamental improvement in living conditions. Many serious problems of course remain and I do not seek to minimise them, but I ask the House to consider the facts. In 1971 nearly 3.2 million dwellings in England were either unfit, lacked at least one basic amenity such as a

> bathroom, or were in serious disrepair. By 1981 that figure had been reduced to just over two million, of which 300,000 were not being lived in. Therefore it is misleading to argue that we are facing a situation of crisis proportions (4) (author's emphasis)

This is the officially received view. However to establish trends the EHCS report compares the 1981 figures with revised figures for 1976, rather than the values established by the 1976 survey. This was done because the 1981 survey used different (and, the DOE argued, improved) techniques for assessing unfitness and repair costs, and in order to compare over time it was necessary to rework the results of earlier surveys on the basis of 1981 survey methods. The result of this revaluation was that trends in unfitness and disrepair appeared more favourable and the EHCS became the subject of some controversy. (5) The statistics in table 1 show that the extent of unfitness was apparently underestimated by over 40% in 1976. Thus instead of a sharp and substantial increase in unfitness, the trend in the late 1970s was that of a marginal improvement.

The trend in disrepair as presented in the survey was rapidly upwards - a 22% increase in five years in dwellings needing repairs of more than £7000 at 1981 prices. However, no information was provided on trends at lower levels of repair costs. The reason given was the unreliability of the 1976 data for these lower levels of repair costs, which would make comparisons misleading. But critics pointed out that the disrepair indicator most often used from the 1976 survey is that of fit dwellings needing more than £2340 (at 1976 prices), involving 910,000 dwellings. At 1981 prices this figure is approximately £4500 and the numbers shown by the 1981 survey to be in this category are 1,312,000 - an increase of 44% in five years. (6) This threshold exceeded the limits of eligible expense for a Repair Grant (£4000 in 1981) and could therefore reasonably be argued to be the appropriate measure of 'substantial disrepair'. If it had been possible to use the £4500 threshold as the indicator for comparative purposes instead of the £7000 used in Figure 1, the overall picture would have been one of no improvement in conditions in the period 1976-81.

Whatever the precise trends in the period 1976-1981, there was clearly no room for complacency when accelerating rates of disrepair had generated a £30,000 million backlog of work (the cost of bringing

Table 1. Changes in Dwelling Conditions 1976-81 (England)
(000s of dwellings)

	1976 published figure	1976 adjusted figure	1981 figure	1976-81 change (%) (unadjusted)	1976-81 change (%) (adjusted)
Dwellings lacking one or more basic amenity	1,531	not applicable	910	-41	not applicable
Unfit dwellings	794	1,162	1,116	+41	-4
Fit but lacking one or more basic amenity	921	746	390	-58	-48
All dwellings needing repairs over £7,000	not available	859	1,049	-	+22
Fit but needing repairs over £7,000	not available	395	574	-	+45

Sources: DoE (1978) English House Condition Survey 1976 Part 1 Report of the Physical Condition Survey, HMSO DoE (1982) English House Condition Survey 1981 Part 1 Report of the Physical Condition Survey, HMSO

Figure 1 Changes in defective housing 1971, 1976, 1981

Thousand dwellings

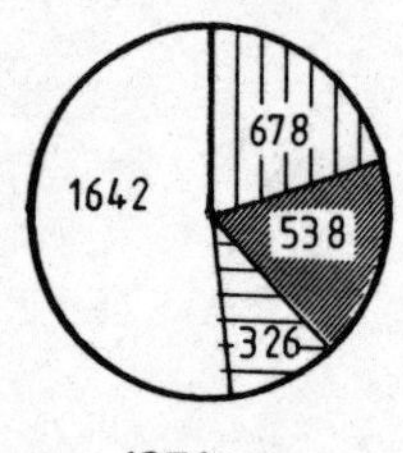

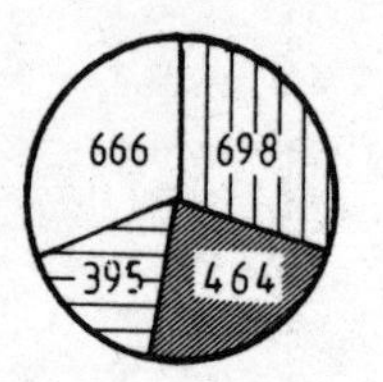

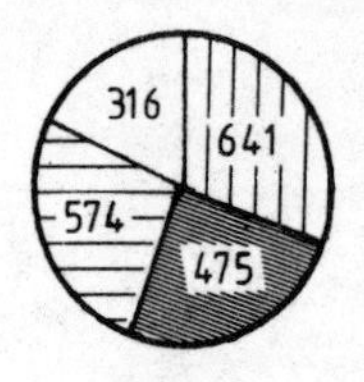

1971
Total 3184

1976
Total 2223

1981
Total 2006

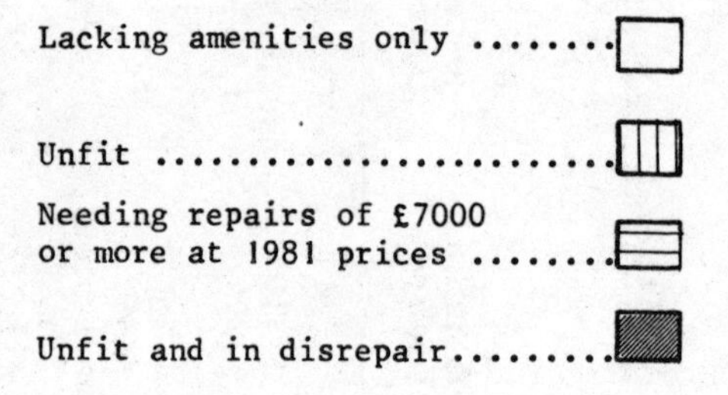

Source: DOE (1982) English House Condition Survey 1981. Part I. Report of the Physical Condition Survey. HMSO.

all dwellings up to the ten point standard) with a quarter of all dwellings needing more than £2500 worth of repairs.

Changing Renewal Problems

Changes in the volume of defective housing are obviously important indicators, but it is the changing composition of problems which reveals the crisis looming ahead. An alternative way of looking at Figure 1 is to focus on the houses in worst condition - the hatched areas. The levels changed only slightly between 1971 (1.54 million) and 1976 (1.56 million), but rose sharply to 1.69 million in 1981. The significance of this expanding core of the worst housing was highlighted in EHCS Part 2 report. Not only had serious disrepair increased by a fifth, but the nature of the constant level of unfitness had changed as demolition had progressed. The stock of unfit houses is increasingly dominated by disrepair and less by design factors, (internal arrangement, sanitary fittings, internal fittings etc). But defective designs are finite in number, whereas decay into unfitness through disrepair is a dynamic and cumulative process. The report states unequivocally that 'the scope for further impact on the worst housing hinges therefore on the extent to which disrepair is tackled' (para 2.21). The EHCS Part 2 report also demonstrated significant changes in what it termed 'the social profile of the unsatisfactory stock', changes which are central to the development of effective policies. In the past population turnover had been an important factor in private sector renovation, involving younger and relatively more affluent households buying cheap houses and immediately modernising them. However, the 1981 survey identified a changing picture in the late 1970s, with the rate of turnover of lower quality housing now decreasing:

> the worst stratum is increasingly the province of long-term residents with limited economic resources or opportunities; with a propensity to accept or tolerate their conditions; and with little or no expectation to move. It was also found that dwellings requiring the greatest expenditure tended to contain the households with the lowest incomes. (para 2.22)

The changes in the social profile are quantified

in Table 2. There was a sharp increase in the proportion of households with nobody in paid employment. The higher proportion of long term residents was not matched by an increase in the proportion of the elderly, indicating that the population occupying the poorer stock is becoming more static. In the early 1970s the private rented sector dominated poor quality housing, but the 1976 EHCS revealed emerging problems in the owner occupied stock. By 1981 the owner occupied sector contained more houses in poor condition than the private rented sector, although it was in the latter where the problems were proportionately the greatest, particularly the continued absence of standard amenities.

Thus the changing nature of physical renewal problems - the increasing dominance of decay caused by underinvestment in repair and maintenance - has been associated with a change in the social role of older housing. The expanding core of bad private housing is increasingly accommodating poor tenants and owners - the unemployed, the elderly and others on low incomes - on a long term rather than transient basis.

Few occupants of the worst strata of dwellings regarded their houses to be in bad condition. Few could afford to finance the necessary remedial works from savings or income - three-quarters of the households in houses unfit or in serious disrepair would have to find a sum greater than annual in-income. Even if they received a grant many would find difficulty servicing a loan to cover their share of the costs. In sharp contrast, many households in the somewhat better quality unsatisfactory stock could finance remedial works from income, savings and loans - about half those eligible for repair grants and a third of those eligible for improvement grants.

The expanding core of bad housing is the leading edge of urban decay and is the most serious consequence of the failure of renewal policies in the 1970s.

It reflects a reduction of the clearance programme from 64,000 p.a. in 1973 to 38,000 in 1981, together with the inability of contracting renovation programmes to stem the rising tide of substantial disrepair, despite the improved targeting of a higher proportion of improvement activity on people and areas in greatest need after the Housing Act 1974. The EHCS evidence on the impact of renewal programmes showed that in areas of poor housing with a static, low income, population, the concentration of public

Table 2: Changes in the Occupants of Unsatisfactory Dwellings 1976-1981 (England) (% of households)

	Satisfactory Dwellings	Fit but in Serious Disrepair		Unfit Dwellings		Lacking a Bath	
	1981	1976	1981	1976	1981	1976	1981
Single person 60+	14	14	16	26	25	34	29
Older smaller households	18	24	19	22	20	25	31
Household head retired	22	26	27	36	33	39	47
No household member in paid employment	29	20	35	30	47	31	63
Resident for over 20 years	25	29	32	33	48	36	55
Owner occupiers	59	49	59	35	53	29	38

Source: DOE (1983) English House Condition Survey 1981 Part 2 Report of the Interview and Local Authority Survey HMSO

investment in HAAs and GIAs had been sufficient to hold serious disrepair and unfitness steady. But in these types of areas which had not been designated, but where the potential for area action existed, the worst conditions had increased.

The bald statistics of the 1981 EHCS suggest that problems of bad conditions are still overwhelmingly concentrated in the older private stock. Of the 4.3 million properties requiring more than £2500 worth of repairs, approximately half are owner occupied, a quarter private rented and a quarter were public rented (see also Table 3). However, a comparison between tenures of this repair backlog on the basis of age of property (Table 4) shows that not only were problems emerging in the interwar and postwar stock, but that post 1919 public sector housing is clearly in a worse state of repair than owner occupied housing of the same age. (7)

Structural defects in system built estates are often compounded by unacceptable environmental conditions resulting from inappropriate design. A large scale study of the relationship between certain design features and the incidence of social problems has recently generated systematic data about the conditions facing future renewal programmes. (8) Furthermore, the underlying causes of deprivation among the two million or so people living in Outer Housing Estates - the overspill areas of the 1950s and 1960s - have been analysed by the CES Outer Estates Project. This work demonstrates that the inadequate state of repair of much of the housing is combined with low incomes and inaccessibility to jobs and services to produce Britain's neglected areas of urban deprivation. (9) The 300,000 'difficult to let' council houses are but the tip of the iceberg. The Association of Metropolitan Authorities estimate the total repair bill for all types of council housing at £19,000 million (see chapter 3 above).

The Building Employers Confederation estimates that the cost of repairing private sector housing would be £20,000 million, whereas the AMA estimate is £25,000. (10) This gives a total backlog of some £39,000 million - £44,000 million worth of work, now widely agreed to be a more accurate estimate of the scale of accumulating problems than the £30,000 million estimated by the EHCS.

In sum, the weight of the available evidence points to a failure to deal effectively with the 'traditional' renewal problems of old private housing in the 1970s being compounded by the emergence of new problems in the public sector. Yesterday's

Table 3: Housing Condition by Tenure 1981 (England) (000s of dwellings)

	Unfit		Substantial Disrepair		Lacking Basic Amenities	
	No.	%	No.	%	No.	%
Owner occupied	493	44	539	51	340	37
Local authority of new town	67	6	50	5	143	16
Rented from private landlord	370	33	343	33	282	31
Vacant	196	18	117	11	145	16
All tenures	1,116	100	1,049	100	910	100

Source: DOE (1981) English National House Condition Survey Part 1 Report of The Physical Condition Survey HMSO

problems have only been partially resolved and yesterdays's solutions have become part of today's problems. The result is the emergence of an expanding core of bad housing occurring across tenures, accommodating increasing concentrations of poor owners and tenants.

Table 4: Percentage of Dwellings in Need of £2,500 + Repairs 1981 (England)

	Owner Occupied	Public Rental	Private Rental
Pre 1919	49	46	56
1919-1944	17	21	33
1945 +	3	8	3

Source: DOE (1981) English National House Conditions Survey Part 1 Report of Physical Condition Survey HMSO

POLICIES AND OUTCOMES IN THE EARLY 1980s

These problems point to the need for policies which would provide or stimulate significantly higher levels of investment in replacement, repair and improvement than had been the case in the late 1970s. Renewal strategies would need both remedial and preventative components, targeting investment on the circumstances producing an expanding core of bad housing in both public and private sectors.

A recent analysis used EHCS data to estimate the underlying rate of deterioration (i.e. net of the effects of clearance and grant aided improvement) as 115,000 dwellings per annum in 1976-81. Some 75,000 were falling into the £7,000 plus repair band and 40,000 becoming unfit, an underlying trend which continued into the 1980s. (11)

This is almost certainly an underestimate as it did not fully take into account the accelerating deterioration of council housing. Nonetheless, a crude estimate of what is needed to halt the further growth of the core of bad housing would be an annual programme of 115,000 of the worst houses replaced or improved, together with sufficient investment in the slightly better quality stock to prevent an acceleration of the rate of decay into these categories - say 10,000 dwellings a year improved. Building from this, several levels of remedial programmes can be postulated:

(a) a holding operation - requires a targeted programme of 125,000 dwellings a year;
(b) a minimal improvement of housing conditions - elimination of the core of 1.69 million bad houses in 1981 by 2001 - re quires a further 85,000 of the worst houses per year to be dealt with giving a total programme of 210,000 dwellings;
(c) a significant improvement of the housing stock - the elimination of the 1.69 million backlog by 1991 would require 170,000 dwellings per year renewed over and above the holding figure - a targeted programme of 295,000 dwellings per year cleared or improved. (12)

These targets are rudimentary and could be varied to take account of changing circumstances. Targets for clearance with public sector replacement and grant aided renovations could be revised downwards to the extent an increased proportion (compared with 1976-81) of the occupants of cleared properties are able to buy replacement housing and an increased proportion of the worst or slightly better condition properties are improved without grant aid. Moving from a 'worst first' approach, it can be argued that over a ten year programme it would be possible to deal with the underlying rate of deterioration into the worst categories by changing the balance between the remedial and preventative work. A higher proportion of preventative work would, in theory, eventually reduce the rate at which properties fall into the core of worst housing.

Declining Programmes

The impact on housing of right wing ideology and monetarist economics, has been profound. Within these dominant parameters cuts in public expenditure are not only necessary for the implementation of economic policy but also have the virtue of promoting the new order. As chapter two above has shown, housing has borne the brunt of expenditure cuts and in no other area of social policy is the new order as clearly in evidence, centering on the myopic pursuit of increased levels of owner occupation at all costs.

In this context it was hardly surprising that by

the beginning of 1982 Michael Heseltine's 1979 pre-election promise to reverse the disastrous decline in housing improvement under Labour had a distincly hollow ring. New building, slum clearance and renovation had all declined.

Table 5: Housebuilding and Slum Clearance (England) (000s of dwellings)

	(a) Housebuilding Completions			(b) Slum Clearance
	Public	Private	Total	
1979	91.1	118.4	209.8	32.4
1980	93.9	109.0	202.9	28.7
1981	71.5	96.9	168.4	28.3
1982	41.9	104.8	146.7	22.8
1983	43.0	121.3	165.3	20.0 (est)
1984	41.8	131.1	172.9	10.0

Sources: (a) DOE Press Notice 96, 6th March 1985
(b) DOE Press Notice 4, 7th January 1983 and Cmnd. 4513 (1985) Home Improvement - A New Approach HMSO

The impact of these cuts is shown in tables 5 and 6. Council house building output had fallen substantially and was set to fall further as local authorities were forced to massively reduce the volume of new build projects going into the pipeline in order to cope with falling resources. Private house building was certainly not filling the gap as high interest rates and escalating unemployment depressed demand. Clearance rates continued to decline and fewer properties were put into the pipeline as housing authorities anticipated increasing rehousing problems. Renovation of council housing fell by a third, squeezed by cuts in HIP allocations and expenditure committed on new building contracts.

Private sector improvement was to be stimulated by the 1980 Housing Act's modifications to the renovation grant system and the launching of low cost home ownership initiatives. Increased eligibility expense limits, reduced improvement standards, the wider availability of repair grants and the extension of the highest rates of grant to all 'priority

Table 6: Renovations of Dwellings with the Aid of Grant or Subsidy - (England) (000s of dwellings)

	PUBLIC SECTOR			GRANTS PAID TO PRIVATE OWNERS AND TENANTS				
	Local Authorities	Housing Associations	All	Conversion & Improvement	Intermediate & Special	Repair	All	All Renovations
1979	76.0	17.2	93.2	57.2	7.8	0.3	65.4	158.5
1980	77.3	14.8	92.1	65.8	8.1	0.5	74.5	166.6
1981	52.9	11.3	64.2	49.1	14.7	5.1	68.9	133.2
1982	57.7	17.3	75.0	54.7	20.6	28.7	104.0	179.0
1983	85.5	14.4	99.9	79.5	27.2	113.1	218.0	319.7
1984	86.8	18.6	105.4	84.3	29.1	116.2	229.6	335.0

Source: DOE Press Notice 96, 6th March 1985

cases' (i.e. for poor quality houses whether or not they are in an HAA) were included in the package, together with a subsidy for environmental works for HAAs on the same basis as GIAs. The virtual halting of municipalisation and pressure on local authorities to sell unimproved houses was the corollary of launching Improvement for Sale (IFS) and shared ownership schemes. But all these changes had little immediate impact. The programme stagnated, badly affected by the 1980 moratorium, as many councils responded by stopping or reducing discretionary grant approvals in order to meet committed expenditure on new build.

The housing associations contribution to renewal was similarly reduced by precipitate cuts in resources. A moratorium halfway through 1980/81 brought an abrupt end to the rapid expansion of this sector since the mid-1970s. Inner city housing rehabilitation programmes were particularly badly affected as housing association production of renovated dwellings fell by a third. In 1981/82 annual cash limits were imposed on the Housing Corporation and implemented through the introduction of the Approved Development Programmes (ADP). This established direct ministerial control over the volume and composition of expenditure, paving the way for a reorientation of housing associations' activity in line with government priorities.

Thus by the beginning of 1982 the total programme, clearance plus renovation, was running at 133,000 units a year compared with 190,000 unit level inherited in 1979 and a targeted 125,000 postulated for a holding operation. This lower programme was more reliant on voluntary improvement by private owners, a shift which meant a reduced impact on private rented and owner occupied housing in the worst condition. (13) Overall housing conditions were almost certainly deteriorating.

Clearance Slump - Improvement Boom

The period 1982-84 saw an upturn in public sector housing investment as reductions in HIP allocations were offset by the reinvestment of capital receipts and the housing association movement successfully campaigned for a significant restoration of its programme. But during this period the local authority investment programme became increasingly dependent on capital receipts and the distribution of investment between tenures and type of work was completely restructured. The collapse of clearance to residual levels was paralleled by a spectacular boom in improvement, producing the highest ever annual total of renovations in 1984, heavily skewed in favour of grant aided improvement by private owners.

In 1982 the government moved from stop to go in its management of overall housing capital expenditure, urging local authorities to reinvest their capital receipts, increasing the HIP allocation and authorising unlimited expenditure on improvement grants until the end of the financial year 1982/83. The result was that spending was on target, but net capital expenditure (outturn expenditure minus capital receipts) shrank to a third of its 1979/80 level in cash terms, because of the surge in capital receipts. For the past two years local authority housing capital expenditure has significantly exceeded government targets. By 1983/84 local authorities had geared up to higher expenditure levels, were subject to pressures from improvement grant commitments entered into during the previous year, and they legally 'overspent' by using accumulated capital receipts. Despite exhortations for a 'voluntary moratorium' in July 1984 and a reduction of the prescribed proportion of the year's capital receipts from 50% to 40%, the 'overspend' in 1984/85 was even higher at an estimated £500 million.

As a result of the pipeline effect, council house building declined precipitately in 1982. Whilst private sector construction recovered slightly it failed to fill the gap and overall house building in the years 1979-84 fell half a million units short of the needs estimated by the 1977 Green Paper, the last official national estimate. Councils' housing resources were further restricted by reduced relets because of sales. With increased pressure on reduced rehousing resources, the slum clearance programme collapsed to a residual level of 10,000 units per annum in 1984.

However, clearance emerged as a component of the <u>renewal of council housing</u>, and in the early 1980s over 10,000 were demolished, despite leaving debt charges to be paid for a further 45 years on average. The proportion of investment allocated to renovation increased substantially at the expense of new construction, a not unexpected outcome given that local authorities have a particular responsibility to existing tenants. Restructured local authority investment programmes produced a recovery of council housing renovations to a level slightly above that of 1979 (table 6).

But these renovation programmes were totally inadequate in the face of the burgeoning scale of known defects, especially when limited resources were being diverted from routine modernisation to cope with emergency work on industrialised units. Despite growing publicity about the scale of the crisis threatening to overwhelm some authorities, the government refused requests for additional resources.(14) The only concession was the Housing Defects Act 1984 which provided substantial assistance to private owners who had bought defective council houses. Instead the government pursued the twin themes of better use of existing public resources and involving private capital in estate renewal.

The well documented Priority Estate Projects (15) (PEP) have demonstrated that the problems of a combination of inadequate and unpopular designs, lack of facilities, poor management and maintenance, vandalism and voids, can be more successfully confronted through innovations in housing management and imaginative physical improvements which use resources more effectively. In autumn 1983 these experimental projects were given an extension to 1986/87.

The second theme has been pursued with increasing vigour through the development of various partnership arrangements between local authorities and developers. A prominent innovation has been the renovation since 1982 of over 20,000 units of former council accommodation bought in blocks by property speculators from councils, improved and then sold to individual owners. (16) Hence 'difficult to let' estates which were clearly 'impossible to sell' on an individual basis, have proved possible to sell on block basis for refurbishment (though often at higher discounts) as part of a secondary round of privatisation. In many cases these renovation schemes are only commercially viable with a substantial subsidy, such as an Urban Development Grant, to cover the gap between renovation costs (including the developers profit) and the expected return. In return for the discounts on initial sale to the developer and subsidy for renovation, local authorities get the right to nominate tenants from the waiting list to be given first chance to buy a proportion of the renovated units.

The modest recovery of council housing renewal has been dwarfed by the boom in <u>private sector housing improvement</u>. In the run up to the 1983 general election the government was able to engineer this boom through interlocking changes in housing subsidy systems. The radical change to a deficit funding subsidy for housing revenue accounts introduced by the Housing Act 1980 meant that by 1982 the majority of housing authorities were 'out of subsidy' and expenditure of capital allocations on council housing attracted little or no Exchequer contribution. In contrast, when the government decided to 'turn on the taps' in 1982, it did so in a way which directed the flow of public investment into private housing through specific Exchequer contributions (up to 95%) which enabled local authorities to sharply increase spending on grants at very little cost to the ratepayers. As the 1985 Green Paper stated, with disarming frankness:

> ...these contributions have been used as a method of influencing local authorities spending patterns, particularly by increasing the level of support towards home improvement grants. (17)

The private sector improvement boom had its origins in a series of events in 1981. Riots in the inner cities briefly brought urban renewal back to the centre of the political stage. Part of the political response was the establishment of the Financial Institution Group (FIG), comprising secondees from building societies, banks, pension funds etc, with a brief to identify ways of increasing the private sector contribution to inner area revitalisation, including housing renewal. In parallel, the 1981 EHCS was underway, setting a timetable for a future debate about the condition of the nation's housing stock. Under pressure to make enveloping eligible for subsidy as part of main housing programmes, the DOE decided to undertake an evaluation of the schemes carried out in Birmingham and other provincial cities using uban programme funds. In quick succession a series of independent reports were published by professional institutes, local authority organisations and pressure groups, all drawing attention to the decline in renewal activity, the case for more resources and for a reform of the absurdly complex system for the renewal of private housing. (18)

The reports were followed through in early 1982 as delegations presented their arguments to senior civil servants. Increasingly, and more significantly, construction industry interests were pressing for a halt to the downward trend in public sector capital investment. By this time the preliminary results of the EHCS were available within the DOE and were reported to be causing considerable ministerial concern. In anticipation of the adverse publicity which publication of the results would generate, civil servants were identifying possible short term initiatives in the context of an internal review of longer term policy.

The first move came when the spring budget provided a minor stimulus to the building industry. This included the provision that for applications received before the end of 1982 for intermediate and repair grants, the rate of grant would be increased from 75% to 90% of eligible cost, with an Exchequer contribution of 95%. When the EHCS results were announced in December the government was able to recite a litany of initiatives which it had by then taken to address renewal problems: successive time extensions for the 90% grants, first to the end of 1982/83 and then to April 1984; increased capital allocations for grant expenditure; the launching of enveloping as a part of national policy; and the establishment of a DOE - Local Authorities Association Working Group on Home Improvement - to consider short term policy changes in the light of the EHCS results and to report by March 1983. The politics of improvement policy were clear by early 1983 when John Stanley, Minister for Housing and Construction said:

> As we start 1983, all the signs are that both new house building and home improvements will be spearheads of economic recovery this year...a veritable explosion in home improvement is now taking place, with expenditure on improvement more than doubling in the course of just one year. (19)

The boom was in full swing during the election and peaked in 1984 when 230,000 grants were paid to private owners and tenants.

This boom was the product of expediency and opportunism. A one-off concession to the construction industry released pent up demand for grants by making them particularly attractive to both house owners and local authorities. The unexpectedly rapid increase in applications in the context of an approaching general election was too good an opportunity to miss. Expediency prevailed over Treasury orthodoxy, enabling the government to claim that it was dealing effectively with the situation revealed by the EHCS. Moreover, when 'underspend' was succeeded by 'overspend' in 1983 (as local authorities supported higher programmes by capital receipts) the bitter pill of breached cash limits was sugared by the improvement boom channelling the surge of local authority capital expenditure into private housing, largely at the expense of council house building. By 1984 the proportion of local authority capital programmes taken up by private sector improvement had increased fourfold, from 6% in 1980/81 to 24% in 1983/84.

There was nothing fortuitous about central government direction of housing association programmes following the introduction of the ADP system in 1981/82. As local authority HIP allocations were reduced, funding of housing associations by local authorities was an early casualty and by 1983 Housing Corporation finance accounted for 87% of housing association programmes. Partly because of housing associations campaigning, but mainly as a result of government's wish to marginally increase public spending in 1982 in a sector which it could control, the Housing Corporation's outturn expenditure was restored from its low of £520 million in 1981/82 to £775 million in 1982/83. The total was subsequently cut back to £687 million by 1984/85 as the government sought to contain overall public sector housing investment. Compared with local authorities, capital receipts have been far less significant and the Housing Corporation's outturn figures have been closely in line with government targets, but at a significant cost to individual associations in terms of disruption and perennial uncertainty. (20)

Up to 1981 the main priority for housing association investment was providing rented accommodation through inner city projects and building for special needs. This investment programme has been restructured in line with government priorities by annual increases in the proportion of the ADP allocated to subsidised schemes for sale. Low income home ownership has been promoted by Improvement for Sale, Shared Ownership, and Leasehold Schemes for the Elderly. Although production to date totals only 25,000 units, by 1983/84 home ownership projects accounted for a third of the units going into the development pipeline. (21) This shift has been mainly at the expense of inner city rehabilitation schemes. The production of fair rent units in inner areas fell by 50% between 1979/80 and 1983/84, from 20,000 to 10,000 per year. To cushion themselves from the effects of uncertain allocations or to avoid toeing the government's line on low cost home ownership initiatives, some associations have diversified: increasing involvement in agency services for community groups involved in environmental improvement, or for elderly owner occupier improvers, are potentially significant innovations. (22)

The Impact of the Boom

The impact of the expansion and restructuring of the housing renewal programme since 1982 can be broadly assessed in relation to the following issues: the impact on the housing stock conditions identified by the EHCS in 1981; the social distribution of the benefits of the boom; and the politics of policy.

Superficially, it could be argued that in relation to estimated thresholds of 210,000 dwellings per annum cleared or renovated to achieve minimum improvement in housing conditions and 295,000 a year for significant improvement, the three years 1982-84 produced a substantial step forward, with successive outputs (clearance plus renovations) of 202,000, 328,000 and a peak in 1984 of 345,000 units. However, if one goes beyond the aggregate volume of output to examine the composition of the boom, this assessment must be heavily qualified.

The boom has been dominated by low cost private sector improvements (see table 6) through intermediate and repair grants. However, the government's Distribution of Grants Enquiry (DGE) showed that in the case of owner occupiers it was improvement grants which were concentrated on the worst houses whereas repair grants and intermediate grants went to dwellings in relatively better conditions. (23) From the DGE data it seems reasonable to add a half of the repairs and intermediate grants totals to the total of improvement grants paid to give an estimate of the impact of private renovation grants on houses in, or about to fall into, the core of worst housing. Supplementing these totals by the totals of slum clearance and Housing Association renovations together with a half of local authorities' renovations, gives a generous estimate of the overall impact. Applying these assumptions to the statistics in tables 5 and 6, the totals for 1982-84 are, successively 148,000, 219,000 and 221,000, - an annual average of 196,000. Thus on these generous assumptions a minimum improvement of housing conditions has been achieved in the past three years which, if sustained, would eliminate the core of bad housing by early in the next century. However, as will be discussed later, this will not be sustained beyond 1985.

If the 1986 EHCS uses the same methods as in 1981, the results will probably show that despite the improvement boom, very little progress was made in eradicating the core of bad housing in the early 1980s. However, if it more effectively measures the condition of council housing, it will almost certainly show that the core of bad housing continued to expand.

These pessimistic predicitons are given further credence by a consideration of other factors relevant to the impact of renewal programmes. The rapid expansion of grant aided work was not accompanied by increases in staffing, with the result that proper supervision of grant work was extremely difficult. This, combined with the reemergence of 'cowboy builders' as the building industry overheated, has meant a lot of dubious quality work, poor value for money and, in the case of the poorer quality property, the likely need for early remedial reimprovement.

In contrast there has been no boom in enveloping, the only recent initiative with the potential for dealing with poor housing, through quality controlled comprehensive improvement of the external fabric of whole blocks of houses. Enveloping was heralded by Housing Minister John Stanley, as equipping local authorities with "...much the swiftest instrument they have ever had..." (24) for housing improvement - it was then comprehensively tied up in the red tape of scheme by scheme Treasury approvals. (25) Recent changes have by no means fully resolved these problems and with a total of only 5,600 dwellings included in 60 schemes in England, the potential of coordinated improvement and repair schemes has simply not been realised. (26)

An appraisal of a housing improvement boom engineered primarily on the basis of incentives for voluntary private sector improvements inevitably raises the question of who benefits. The DGE addressed this issue, but the results published to date are somewhat limited. Whilst the proportion of professional and managerial owner occupiers who received grants was smaller than the proportion who were eligible for assistance, the fact remains that 13% of grants went to those affluent groups. In contrast, owner occupiers in the partly skilled or unskilled categories received only 20% of the grants. (27) In the public sector the reorientation of the housing association programme from fair rent accommodation to low income home ownership initiatives has possibly not adversely affected the number of bad houses renovated, but has certainly meant that a lower proportion of resources has gone to people in greatest need. Similarly, the sale for refurbishment of council estates has achieved the initial upgrading of bad housing for a new class of owner occupier. But the people decanted from the blocks prior to refurbishment, or those on the waiting list who cannot afford to buy, do not benefit from this publicly subsidised speculative development, especially if the councils cannot invest the capital receipt.

In sum the expanded programmes have resulted in marginal improvement to the overall condition of the nation's housing stock, but the emphasis on owner occupation has meant that compared with the late 1970s a smaller proportion of the available public resources went to those in greatest need. There is clearly scope for improved targeting of resources. Even this expanded programme is not really getting to grips with the fact that the worst houses are owned or tenanted by those least able to invest in improvement.

In terms of the politics of policy the government achieved strong centralised direction of the overall renewal programme and was able to reshape it in line with the priority for home ownership. There has been a substantial shift in public capital investment programmes (both local authority and Housing Corporation) away from new build for rent to improvement by or for owner occupiers. This restructuring during the period 1979-84 can be interpreted as taking housing renewal well down the road to privatisation, by substantially shifting the thrust of public investment to supporting owner occupation. However, the interlocking effects of expenditure policy, the emphasis on home ownership and political expediency generated contradictions with important implications for the future.

The capital receipts system, introduced as an incentive for local authorities to sell council houses, (28) effectively deprived the Treasury of its control over the volume of housing capital investment, but the net provision (new money) was reduced to a small proportion of its 1979/80 level. The improvement boom channelled this surge of expenditure, but whilst this was politically expedient it resulted in relatively high levels of programmes. However, the effect has been to shift a greatly increased (6% to 24%) proportion of local authority investment into a spending area which is vulnerable to criticism on the grounds of not being accurately targeted on needs. Finally, the boom in private sector improvement was paralleled by the widespread publicity given to the growing body of evidence of a 'house condition time bomb' ticking away in the council housing sector.

PROSPECTS FOR THE 1990s

As the second Thatcher administration approaches mid-term it is becoming clear how the contraditions in housing renewal policy will be resolved. The road to full privatisation is signposted by the crisis in the public sector housing capital programme, the 1985 Green Paper proposals for private sector housing improvement, the strategic options delineated in the Housing Corporation's first corporate plan, and the launching of the DOE's Urban Housing Renewal Unit to promote recent initiatives for dealing with run down council estates. The intention is to increase yet further the reliance on private sector solutions to housing renewal problems, by the maximum possible substitution of private spending (by individuals and financial institutions) for public capital investment, linked to further extensions of home ownership. National policy will soon be at a crossroads, but there is an alternative route into the 1990s.

Public Sector Investment Crisis

Immediately after the 1983 election the Treasury's reluctant tolerance of higher levels of local authority housing investment programmes evaporated and the priority for establishing effective cash limits became overwhelming. The failure of the 'voluntary moratorium' on the expenditure of accumulated receipts in 1984 prompted further restrictions. This financial year (1985/86) saw the HIP allocation reduced to £1,650 million (with some £400 million held back to safeguard the cash limit against local authorities 'overspending' their accumulated receipts) and the prescribed proportion reduced to 20%. If these measures are successful, councils' housing capital spending will fall from around the £3000 million mark to about £2300 million. The uncertain effectiveness of these controls is likely to mean early legislation to remove local authorities' opportunities for topping up HIP allocation by a prescribed proportion of capital receipts. (29) From the Treasury point of view the built-in pressure to quickly rein in housing capital spending is enormous. This stems from the overwhelming dependence of existing programmes on capital receipts. Compared with £2862 million net provision ('new money') in 1979/80, only £729 million was allocated this year, the balance of the £2,342 gross provision being capital receipts. As capital receipts fall away the choice will be between allocating substitute new money each year to compensate and thus sustain the (already low) investment levels, or cutting the programme. Within the current perameters of expenditure policies the prospect is one of successive reductions down to below £1000 million per annum.

Privatising Private Improvement: the Conservative Blueprint

This Treasury battle to control and substantially reduce local authority housing investment significantly influenced the timing and content of the 1985 Green Paper. Soon after the election the government announced that the temporary 90% grant provisions would not be extended beyond April 1984 and that the review of long term improvement policy was being resumed. It quickly became known that private sector improvements was a prime target for Treasury cuts. This programme had expanded from £112 million in 1970/80 to £425 million in 1982/83 and peaked at £911 million in 1983/84. Although it was falling back to an estimated £750 million in 1984/85 it was clearly vulnerable to further cuts. (30) Claims that the improvement boom was largely subsidising work which would otherwise have been done by owners from their own resources were used by the Treasury to legitimate reducing the eligibility for 90% grants back to pre 1982 levels. Privately, these claims were also used to support the Treasury's case for government withdrawing completely from subsidising private improvement, thereby substantially reducing the pressure for 'new money' to compensate for falling capital receipts. The DOE, acknowledging the need for a reduced programme wanted to preserve the grants system, but achieve more efficiency and effectiveness in the use of whatever resources it could secure, through improved targeting and simplification of procedures. In defence of the grant system the DOE mounted the Distribution of Grants Enquiries, generating evidence which denied the more extravagent of Treasury claims.

The interdepartmental conflict resulted in an elephantine gestation period for a new policy to go beyond the expediency of 1982-84. (31) Apparently the resolution involved decisions at prime ministerial level. The balance struck between the objectives of reducing expenditure, improving targeting and simplifying procedures was finally embodied in the Green Paper, Home Improvements: A New Approach, published in May 1985. It is now no longer necessary to speculate about the key features of the government's response when, in all probability, the publication of the quinquennial EHCS results in 1987 will coincide with the run up to the next General Election. Either there will have been, or will about to be legislation based on the Green Paper:

> The policies now being established provide the blueprint for improvement policy over the remainder of the century... The cornerstone of policy must be that owners are primarily responsible for the condition of their houses, though they should be given appropriate help and encouragement in shouldering their responsibility. (para 81)

Within this perspective the role of government is reduced to one providing limited means tested financial assistance to the occupants of housing in the worst condition, together with encouraging 'timely private spending' on improvement and repair by those not eligible for assistance. The proposals are summarised in figure 2.

Acknowledging the EHCS evidence that many households have insufficient savings or means to pay interest on a commercial loan the Green Paper states:

> ...the overriding aim will be to ensure that the (home improvement assistance) system includes those who are unable to afford necessary improvement and repair work unaided, and excludes those who do not need help. (para 28)

It is difficult to judge whether the proposals will achieve this because no details are given about the tests of eligibility for assistance, other than that they will be compatible with those for entitlement to housing benefit: this suggests a low threshold of eligibility. For those who pass the means test the form of assistance, grant or loans or both will be determined by whether the property is unfit or not.

Mandatory grants '...will be available only for those dwellings in the very worst condition which are unfit for human habitation and endanger the health or safety of their occupants' (para 36) - subject to them being suitable for renovation. In order to encourage a high rate of take up, wherever possible on a voluntary basis, grants will be available for all work (improvement and repair) necessary to bring houses up to the (new) fitness standard. The grant rates will be a percentage of the cost of eligible works (subject to an upper limit), calculated on a sliding scale which allocates the highest rates to those on lowest incomes in dwellings with the highest cost of works. Thus the new minimum standard of fitness will define both the dwellings eligible for grant and the improvement standard to be achieved. Although the inclusion of some standard amenities is a step forward, this will be more than offset by the exclusion of internal arrangement standards and the restriction of the disrepair criterion to 'dangerous structural disrepair' and the dampness criterion to a condition 'so pervasive as to be a threat to the health of the occupant'. The result would be a lower standard of fitness.

Under these proposals the number of properties eligible for grant could fall by up to two thirds from 2.8 million eligible under the present rules to between 1 million and 1.5 million, depending on the threshold set by means test and the fall in the number of unfit properties through the application of the new standard. Moreover, unless grant rates are at or near 100% it may prove difficult to secure a high take up. Grants will be restricted to a low standard, low cost (possibly £3000-£4000 per dwelling) and short term operation, leaving options open for demolition or reimprovement at a later date. The inherent uncertainty and poor value for money will deter many owners and create conditions which in the past has spawned red-lining by building societies.

The achievement of higher standards depends significantly on the take up of discretionary equity sharing loans which will replace improvement and repair grants. These will be available to poor people living in dwellings which are fit but in need of repair, and they could also be used to 'top up' a mandatory grant (see figure 2). The eligible works would be any combination of repairs and improvements needed to give dwellings a thirty year life, subject to a maximum six point standard (a simplified version of the current ten point standard). Loan amounts as a percentage of the cost of eligible works will be calculated on the same sliding scale as grants. Instead of paying interest the owner will effectively sell a fixed percentage of equity in return for a local authority contribution to the cost of repairs. The percentage would be the amount of the loan as a proportion of the postimprovement value of the property. The Green Paper's example is that of a loan of

FIGURE 2: SUMMARY OF THE 1985 GREEN PAPER PROPOSALS FOR PRIVATE SECTOR HOUSING RENEWAL

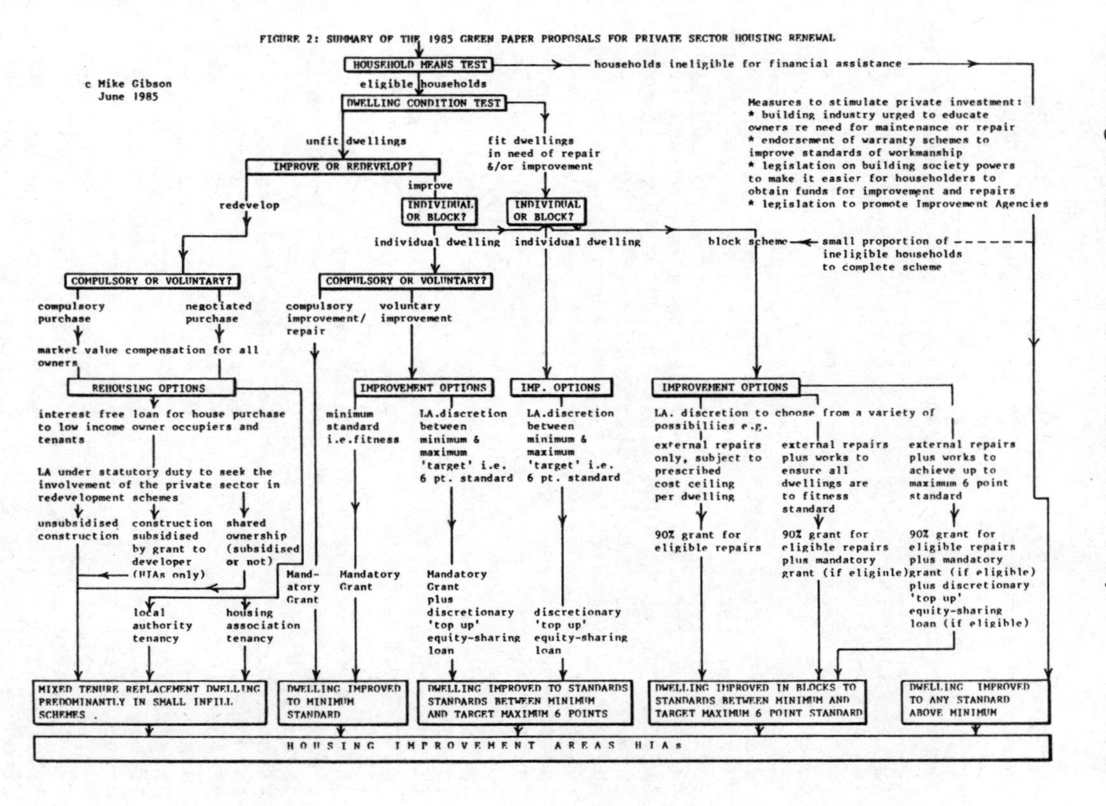

£3,000 on a property worth £30,000 after improvement which would result in a 10% equity share for the local authority. This would be a charge on the property and when the property is eventually sold the owner or executor would pay 10% of the sale price to the local authority: if the house is eventually sold for £50,000, the local authority would receive 10%, i.e. £5,000.

This novel proposal - with a hint of irony, some would call it shared ownership in reverse, or creeping municipalisation - took many people by surprise, especially those working in housing north of Watford, where poor people do not live in houses worth £30,000 and rising rapidly in value. More realistic values in the Midlands and North would be a house bought for £12,000, worth £15,000 after £5,000 worth of loan funded improvement, resulting in a local authority share of 33%, which would involve repaying £6,000 on resale at £18,000 say five years later. The equity share scheme solves the problem of how people who cannot afford to pay interest can be enabled to 'borrow' money without giving them any subsidy. But it falls foul of the valuation gap - the fact that the increase in value of the property after improvement is substantially less than the cost of works, except in the most buoyant of housing markets. This problem of the market undervaluing structural repair and improvement is one of the main reasons for grant subsidy. The prospect of ceding a significant equity share is hardly a comparable incentive. Moreover, without being able to protect their investment through repair and maintenance, it is not immediately obvious that the local authorities will want to acquire what could turn out to be shares in decay. The substitution of loans in this form for grants is likely to substantially reduce the number of fit houses repaired with government assistance.

The proposals for enveloping and block schemes are intended as a response to situations where '...conditions across the whole area are so bad as to discourage individual owners from spending money on their homes'. This is presented as a resolution of the procedural problems which have beset this initiative over the past two and a half years. Individual scheme approvals will not be needed if certain criteria are met. Again the likely effects of the government's proposals are difficult to gauge. The requirement that the great majority of owners should pass the means test may or may not be a major restriction, depending on the threshold set and the way the scheme is administered. The exclusion of curtilage works such as boundary walls will cause problems and prejudice standards as these are expensive items which it will be difficult to persuade owners to have done either with a grant or an equity sharing loan. On balance however, it seems likely that enveloping and block schemes will become relatively more important in local authority renewal strategies because of the possibilities for achieving long term improvements by combining grants and loans (see figure 2).

The proposals for clearance and redevelopment are based on the premises that '...the pendulum has swung too far against demolition' but that future redevelopment should be sensitive, small-scale, mixed tenure and implemented with a minimum of compulsion. The Green Paper stresses the need for swift action against unfit dwellings. Local authorities are to be pressed to regularly inspect their areas and will be required to deal with unfit housing through renovation or closure within a year of the date they are declared unfit. A new code of guidance will set out the social or economic factors to be considered in choosing between improvement or redevelopment. Compulsory improvement and repair powers will be restricted to unfit properties and houses in multiple occupation - a significant reduction of powers, especially if the unfitness standard is reduced.

The reform of clearance and redevelopment is long overdue and several of these proposals (see figure 2) are welcome e.g. improved compensation and other financial assistance for house purchase. The crucial issue is whether the rehousing opportunities will match the needs of the households affected. It is assumed that clearance can be significantly increased without increasing the availability of local authority tenancies, by providing private sector and (although not specifically mentioned in the Green Paper) Housing Association alternatives. The danger is that an assumption of eventual clearance (and therefore no improvement assistance), in the absence of effective rehousing provision, could simply result in blight and accelerated decay. Whilst the emphasis on unfit dwellings is welcome, the resource implications of dealing with them quickly are simply not addressed.

Finally Housing Improvement Areas (HIAs) will replace GIAs and HAAs. What happens in them will largely depend on the use of mandatory grants, equity sharing loans, block schemes and reformed redevelopment programmes. Developers are the only group who will receive special treatment in HIAs. New subsidies for private new build and renovation (via a variant of Improvement for Sale) will be available in HIAs only (to bridge the developer's valuation gap). A single type of statutory improvement area has been included on many agendas for reform since the early 1980s. However, although the Green Paper endorses the importance of area improvement, it is difficult to be other than pessimistic about the future of current area renewal strategies, given the reduced availability of the grants on which they critically depend and the problems local authorities will face in concentrating resources when most of these grants will be mandatory and thus available irrespective of location. Much indeed will depend on realising the potential of block schemes.

Overall the proposals clearly meet the government's Treasury dominated objectives. The substantial reduction of eligibility for government assistance by means testing, the replacement of a grant based system of improvement by one largely dependent on loans, and an emphasis on the role of the private sector in redevelopment, together constitute the full application of the Thatcher administration's privatisation strategy to private sector housing renewal. The proposed system has the capacity to reduce public expenditure in this area to a minimum, depending on the threshold set for the means test and the effect of the proposed standard of fitness. The proposals will certainly target assistance on those in greatest need, but at best will only provide very low standards. The Green Paper is a recipe for deferred demolition in the worst areas and continued underinvestment in the marginally better areas. Elsewhere, the proposals to encourage investment by those excluded from assistance (figure 2) are unlikely to have a measurable effect in increasing the proportion of money spent on structural repairs. The objective of simplification is less fully achieved, as the simplicity of the unitary grant and area improvement schemes is largely cancelled out by the complexities of equity sharing loans.

Fundamentally, however, there is no indication of the likely impact of this new approach on the problems revealed by the EHCS - and conception of a strategy and programme for the renewal of older housing areas is conspicuous by its absence. The proposals are intended to specify the most beneficial application of whatever resources local authorities make available from their annual allocations. If the tide of disrepair is not turned it will presumably be the responsibility of the local authorities for allocating an inadequate proportion of their shrinking resources to priming the pump of private spending.

Public Sector Renewal

Also conspicuous by omission is any significant reference to the role of housing associations. This may well herald the end of an era, albeit a brief one, with the reduction of the once pivotal housing association acquisisiton and improvement programmes to a residual holding operation. In parallel with the finalisation of the Green Paper proposals, the Housing Corporation was preparing its first Corporate Plan. This assessed the consequences of a range of future resource levels, providing a systematic basis for government decision making about the future housing capital programme and the Housing Corporation's share of it.

The Corporation states that as a result of the reduction in the past five years:

> ...our judgement is that in many inner city areas the level is now too low to safeguard previous investment...further reductions in the overall size of the programme would mean diverting resources to the inner cities from other programmes, or concentrating available funds in fewer areas...we do not want to abandon the substantial amounts of past investment in partly completed areas. (32)

The outline of investment options makes it clear that the combination of a reduced Corporation programme and cuts in the availability of improvement grants and other local authority capital investment in designated areas, such as indicated in the government's current expenditure plans, would immediately present ministers with difficult decisions. Concentration would be essential and this would mean a choice between 'the inner city areas and the rest' - the rest being housing for the elderly and other special groups in the outer metropolitan areas and the shire counties.

As in recent years, future reductions in Corporation funding may not be as severe as those imposed on local authorities and may be partially offset by recycling public funds through the refinancing of existing loans from private sources. But the logic of current expenditure policies suggests that there will be cuts in government support. Current parameters of policy also dictate that low cost home ownership schemes will be shielded from these cuts. The prospects for older (particularly private rented) property outside designated areas is therefore bleak. Future investment by associations will become increasingly selective within inner areas, reinforcing past investment rather than extending into additional neighbourhoods. There is certainly no prospect of housing associations being able to compensate for widespread reduction in private sector improvement activity.

The prospects for council housing on this scenario are possibly worse. With understandable reluctance the government has yet to fully acknowledge the scale of defects in public sector housing, and has put in hand its own investigations through the Building Research Establishment. Further substantial reductions in investment will result in an evergrowing backlog of routine maintenance. No matter how imaginative the innovations in management and effective use of resources prompted by the Priority Estates Project or developed from initiative within the housing authorities, they will not compensate for the effects of the investment crisis working its way through. The major political response to the pressure exerted by the AMA since the 1983 General Election has been the increasingly vigorous promotion of various partnership mechanisms for involving private finance in the renewal of council housing. This policy has recently been given a higher profile by the establishment of the Urban Housing Renewal Unit to help and encourage local authorities in these innovative approaches. Clearly this form of renewal should not be dismissed out of hand. But in the context of accumulating underinvestment in the council housing sector overall, its limitations are not only that the benefits are restricted to those who can afford owner occupation, but also the fact that on any conceivable scale of success its impact will be marginal, compared with the withdrawal of public investment. Bereft of subsidies and financially weakened by the sale of better quality housing, the main source of funding for council housing renewal when capital receipts fall away will be rental income from a stock with increasingly high unit costs of maintenance.

Renewal at the Crossroads

The privatisation of housing renewal would be the

culmination of a major phase in the history of state intervention on private housing. This phase would have seen the reestablishment of a minimalist role for central government by a shift from large scale redevelopment to a token level of subsidised clearance, and private improvement. This phase started at the end of the 1960s with 'the partitioning of housing renewal' (33) into a shrinking programme of direct intervention through clearance and municipal rebuilding, and an increasing reliance on voluntary, subsidised, private sector improvement, working indirectly by stimulating the market. The resultant overall downward trend in investment has fluctuated in volume and composition - with brief surges in private improvement in the early 1970s, housing association rehabilitation for rent in the mid 1970s, and private grants in the past three years - but has been apparent since the 1969 Housing Act. (34)

The phase would be completed by the return of renewal to the market place. Government investment would be restricted to minimum levels in a largely self-financing increasingly impoverished and contracting council housing sector, together with a residual level of clearance and minimum standard grant aided private improvement, operated on 19th century 'sanitarian baseline' principles of health and safety.(35) Government housing spending would be dominated by the pursuit of regressive housing finance policies subsidising owner occupation. In more specific terms, from a combination of clearance and mainly full standard subsidised improvements, giving an annual programme at the beginning of the 1970s of some 250,000 units, we would be entering the 1990s with a residual programme of well under 100,000 units, dominated by low standard short term improvements, posing increasing questions about the future of the neighbourhoods in which they are located. It would be possible to celebrate the centenary of the 1890 Housing of the Working Classes Act by a return to a level of state intervention in some respects recognisably similar to that produced by the Act - with the important exception of the subsidisation of owner occupation.

The growing credibility of this scenario means that housing renewal will soon be at a crossroads. Central government will _either_ continue to tolerate a widening gap between the housing standards of the majority and those of a substantial and growing minority, _or_ will have to face the challenge of constucting an effective renewal programme. But surely the return of more effective government intervention as we go into the 1990s is inevitable: as a result of either rejecting the by then widely perceived prospective consequences of privatised programmes or in reaction to a short sharp experience of them, if the present political and economic parameters do not change quickly enough to prevent their full implementation. The issue will be whether, not how, a vigorous public programme is constructed.

It is unlikely to be that simple. The challenge facing those who wish to see a 'renewal of renewal' lies in confronting the processes which are producing the widening gap between a comfortable houses majority and a growing minority housed in an expanding core of bad housing in both public and private sectors. (36) The creation of 'the two nations' in housing is rooted in the process of 'marginalisation' - as the current restructuring of the economy is creating a growing group of people who are permanently unemployed or low paid. Moreover, the unskilled and semiskilled workers who are the victims of structural unemployment suffer disproportionately as state intervention is reorientated away from direct provision. 'Commodification' is a process deriving from an ideological commitment to returning housing to the market place, thus reestablishing the link between income and housing standards which previous intervention partially broke. 'Residualisation' is a term which is now often applied to the processes propelling council housing into absolute and relative decline, providing accommodation increasingly exclusively for the less skilled and the unemployed. But residualisation processes are also affecting the owner occupied sector as it extends to include more low income purchasers receiving limited assistance within the regressive tax relief system, thus expanding the group of owners unable to repair their property. Inequalities within owner occupation are becoming increasingly marked as this sector houses a significant proportion of the marginalised poor in the growing core of bad housing.

The operation of these processes produces the widening gap between a comfortably housed majority and a growing minority who are badly housed. The majority are not economically and politically marginalised and are not trapped in the most unpopular council housing with little hope of escape, or in that decaying owner occupied housing which is losing its investment value in real terms: they have a vested interest in perpetuating the inequalities inherent in the present system of housing finance, which continually diverts vast amounts of public money from housing production into supporting consumption in the private sector. Acknowledgement of this fact means that housing renewal will not automatically be revived simply because there is a growing realisation of the decay of the housing stock.

However, if there is the political will to confront these processes there is no shortage of ideas for the development of a reinvigorated housing renewal programme. Reform of housing finance, increasingly debated in recent years, would be an essential precondition by providing the investment. Another ingredient would be significant changes in the relationship between paid officials and the consumers of housing services through housing management reforms and the further development of participative urban renewal processes: despite (or possibly because of) a hostile environment, reforms, often under the heading of decentralisation, have been gaining momentum. Within this resource and institutional framework local renewal strategies could be formulated which varied the mix between public and private sector finance according to conditions in local housing markets, rather than the balance being dogmatically imposed from the centre. Similarly, it will be critical to develop the capacity for determining the appropriate balance between clearance and improvement in a way which enjoys the support of residents, as a result of providing an appropriate range of rehousing opportunities. The balance between area based action and that oriented to individual houses would similarly need to be judged in the light of local circumstances.

The formulation of these strategies would be able to draw selectively on the innovations of recent years. For example, there are welcome reforms proposed in the 1985 Green Paper based on innovation such as enveloping and agency schemes. More importantly the system outlined in figure 2, with important (but easy to introduce) changes to eligibility for grants and the loan provisions, could be transformed into a potent framework for tackling the renewal of older private housing. Most importantly renewal strategies should be developed on a cross tenure basis. Current tenure based policies and the departmental management systems through which they operate are of decreasing relevance, as the fragmentation of tenure patterns in pre-1919 areas is repeated in council estates and as skills in mixing public and private sector resources to produce socially relevant investment programmes become central to future renewal strategies.

Ideas in plenty, but is there the will? The

housing renewal problems of the 1990s will be the product of the restructuring of the economy and the political response to the pressures which this generates. An alternative to the privatisation scenario has to be developed and the power base constructed from which it can be implemented. Either we grapple more effectively with the problem of managing urban decay and change in a way which minimises inequality, or we face the unpredictable consequence of deepening social division.

NOTES AND REFERENCES

(1) DOE, English House Condition Survey 1981 Part 1: Report of Physical Condition Survey, HMSO, 1982; DOE, English House Condition Survey 1981 Part 2: Report of the Interview and Local Authority Survey, HMSO, 1983.

(2) Association of Metropolitan Authorities, Defects in Housing Part 1: Nontraditional Dwellings of the 1940s and 1950s, AMA, 1983; Defects in Housing Part 2: Industrialised and System Built Dwellings of the 1960s and 1970s, AMA, 1984; Defects in Housing Part 3: Repair and Modernisation of Traditional Built Dwellings, AMA, 1985.

(3) See for example Bristol City Council. Bristol House Condition Survey, 1984, summarised in R. Henderson, 'Grant Aid and the Improvement of Housing in Bristol', Housing Review, vol. 34, no. 3, 1985, p. 108, West Midland County Council, West Midlands House Condition Survey, 1982, Leeds City Council Public Works Committee, A Review of the State of Repair of the Leeds Housing Stock, 1984.

(4) Hansard, 16 November 1983, col. 863.

(5) R. Matthews, 'Conditions, Suspicions', Roof, vol. 8, no. 2, pp. 22-4.

(6) D. McCulloch and D. Smith, 'Older Housing Policy: Future Directions', Housing Review, vol. 32, no. 3, pp. 77-9.

(7) This point is stressed by R. Henderson, 'An Alternative View of Housing Statistics', Housing Review, vol. 34, no. 2, pp. 56-8.

(8) A. Coleman, Utopia on Trial, Hilary Shipman, 1985.

(9) CES Ltd., Outer Estates in Britain - Interim Report: Preliminary Comparison of Four Outer Estates, CES Paper 23, 1984.

(10) J. Carvel, £19 billion needed to right council house defects, Guardian, 5 March 1985.

(11) D. McCulloch, 'An Appraisal of Older Housing Prospects', Housing Review, vol. 33, no. 1, pp.

29-30.

(12) For an earlier attempt to assess the scale of renewal programme needed see Royal Town Planning Institute, Renewal of Older Housing Areas: Into the 1980s, RTPI, 1981, pp. 32-6.

(13) See S. Cameron and R. Stewart, 'Area Improvement in Newcastle - Past and Present', Housing Review, vol. 31, no. 6, pp. 216-18.

(14) See for example N. Wallace, 'Scandal or Modern Slums Forces Demolition', Sunday Times, 11 April 1982; M. Davie, 'Great UK Housing Disaster', Sunday Times, 28 August 1983; J. Carvel, 'Council Houses Facing Urgent £10 billion Repairs', Guardian, 2 June 1983, and 'Council Predicts £56 billion Repair Bill for Tower Blocks', Guardian, 7 March 1984.

(15) See for example DOE, Local Housing Management : A Priority Estates Project Survey, HMSO, 1984 and Peptalk, a bi-monthly newsheet.

(16) J. Cunningham, 'The Flat-Broke Way to a Bank Balance', Guardian 18 June 1985, for developer's publicity see, for example, 'The Writing on the Wall - The Challence of the Inner Cities, Barrat.

(17) Cmnd. 9153, Home Improvement: A New Approach, London HMSO, 1985, para. 77.

(18) The reports were: Shelter Housing Advice Centre, Good Housekeeping: An Examination of Housing Repair and Improvement Policy, Shelter, 1981; Royal Town Planning Institute, Renewal of Older Housing Areas: Into the Eighties, RTPI, 1981; The Institute of Environmental Health Officers, Area Improvement, IEHO, 1981; and Association of Metropolitan Authorities, Ruin or Renewal: Choices for Our Ageing Housing, AMA, 1981; a later addition to the list was London Housing Renewal Group, Avoiding the Bulldozer: The Renewal of London's Private Housing, for a summary of the RTPI and IEHO reports see M. Gibson and T. Brunt, 'Our Older Homes: Twin Calls for Action', Housing Review, vol. 31, no. 1, pp. 10-12.

(19) DOE Press Notice 7, 'Housebuilding and Home Improvement are Spearheads of Economic Recovery says Housing Minister', 12 January 1983.

(20) See M. Langstaff, 'The Changing Role of Housing Associations', The Planner, vol. 70, no. 5, pp. 25-6.

(21) The Housing Corporation Report 1983/84, Housing Corporation, 1984, appendix II.

(22) M. Langstaff op. cit.

(23) Cmnd. 9513, op. cit. pp. 18-23.

(24) DOE, Press Notice 384, 'Enveloping - Min-

ister Announces New Government Home Improvement Initiative, 21 October 1982.

(25) See J. Perry, 'What Boom?' Roof, vol. 8, no. 6, pp. 25-7.

(26) Cmnd 9513, op. cit. para 54, see also J. Morton, 'What's in the Envelope', Local Government News, March 1985, pp. 26-7.

(27) Cmnd 9513, op. cit. figure 1 p. 19.

(28) At the Institute of Housing Conference in 1981, John Stanley, Minister of Housing said "The extent to which authorities can finance new housing capital expenditure is going to depend critically on their willingness and their skill in generating capital receipts..."

(29) See P. Somerville, 'Capital Receipts and the HIP System', Housing Review, vol. 34, no. 3, pp. 85-7.

(30) Cmnd 9513, op. cit. para 8.

(31) There are, of course, no documentary sources about this conflict but the grapevine's tremblings were consistent and the Treasury's views have been quoted on the Panorama programme, 'When the Roof Falls In', 24 June 1985.

(32) The Housing Corporation, Corporate Plan 1985, para 7.4.4.

(33) See S. Merrett, State Housing in Britain, RKP, London 1979, ch. 5.

(34) See M. S. Gibson and M. J. Langstaff, An Introduction to Urban Renewal, Hutchinsons, London 1982.

(35) For a discussion of the concept of the 'Sanitarian Baseline' see R. McKie, Housing and the Whitehall Bulldozer, Hobart Paper 52, Institute of Economic Affairs, 1971.

(36) See R. Forrest and A. Murie, 'Residualisation and Council Housing: Aspects of the Changing Social Relations of Housing Tenure', Journal of Social Policy, vol. 12, part 4, pp. 453-68.

Chapter Five

GROWING CRISIS AND CONTRADICTION IN HOME OWNERSHIP

Valerie Karn, John Doling, and Bruce Stafford

In the past home ownership has been regarded as a relatively unproblematic form of tenure. However, this image is not surviving its rapid expansion during a period of recession. Particularly in the last year growing concern about home ownership has been expressed by academics housing professionals and the media. Contradictions and crises are emerging at a number of levels and for a number of different agencies in the home ownership process, notably for government, as the proponent of rapid increase in home ownership and for the building societies as the main agents of that increase. In this chapter we take these two groups in turn and discuss the dilemmas emerging from the growth of home ownership.

Government

For government there is a growing contradiction between an economic policy which stresses public expenditure savings, a housing policy of expanding home ownership and short term electoral advantage. Until the last few years it had appeared to the Thatcher administration that these three were totally compatible. The promotion of home ownership and especially council house sales had proved a major electoral success; the proceeds of sales could be used to reduce the need for local authority borrowing and as a source of Exchequer revenue; and in the longer term, the switch from renting to home ownership was seen as reducing government's role, responsibility and hence expenditure on housing. This view has proved to be very simplistic for several reasons. In particular it ignored the extent to which the fiscal and social security systems underwrite home ownership. The chief elements in this support are tax relief on interest, interest payments for mortgagors on supplementary benefit and rate contributions for those who qualify for housing benefit, notably elderly owners. Apart from the expansion of home ownership, there have been a number of other reasons why expenditure on these items has escalated in the last five years. The first is rising interest rates, the second is rising unemployment and the third rising numbers of elderly owners needing help with rates.

To take the largest item first, in 1984-85 mortgage interest tax relief had reached £3,500 million, (table 1). The cost of relief had risen from £1,450 million in 1979-80 an increase of 141 per cent. Not only is this a large sum of money, greater than the entire amount paid on housing benefit, but unlike housing benefit it is not targetted on those least able to meet their mortgage costs. On the contrary the size of payment is positively correlated with income. Such expenditure is clearly completly contradictory to an economic and social policy of restricting public expenditure to those areas where it is absolutely essential, and of confining personal subsidies to the most needy. More logically the Treasury would support phasing out tax relief and using the proceeds to lower the standard rate of tax or increase tax thresholds. However, the obvious housing policy for a government interested in promoting home ownership would be to use all or part of the proceeds to fund a housing benefit targetted on marginal buyers. The subsidy would then be designed to increase and sustain home ownership rather than, as at present, pay a bonus to those who would be able to buy without it. However, this is where short term political advantage becomes a problem. Withdrawing tax relief or even replacing it with another system is seen as politically disastrous, given the dominance of existing home owners in the electorate, so such a policy is not being entertained. Indeed, even a policy of gradually phasing it out was handicapped by the Government's decision in March 1983 to raise the eligible mortgage ceiling from £25,000 to £30,000. This failure to tackle the logic of reforming tax relief leaves the government with little room for manoeuvre in the introduction of other kinds of targetted help for owner occupation or renting. For example, as we will see later the problem of the repairs and maintenance costs or poorer owners is becoming a major problem which government cannot ignore in the long term. A problem ignored may be a problem solved for one administration but not for its successors.

Table 1: Cost of Mortgage Interest Relief (£m)

	Total
1979-80	1450
1980-81	1960
1981-82	2050
1982-83	2150
1983-84	2750
1984-85	3500

Source: Hansard, Written Answers, col. 1072, 31 Oct. 1984.

The second major item of expenditure on home ownership is the cost to the DHSS of paying the mortgage interest of people on supplementary benefit. In 1979 there were 98,000 mortgagors receiving supplementary benefit; by 1982 this figure had grown to 235,000. (table 2) In 1984/85 it is said to be at least 250,000. Previously the highest peak had been 124,000 during the recession of 1976/77. The cost to the Exchequer of these payments in 1983/84 was about £150 million compared with £50 million in 1979/80. (1) On average unemployed mortgagors cost more for interest payments than unemployed tenants cost for rent. Though rents have risen, so have mortgage payments because of the sharp increase in interest rates.

Whereas short term electoral interests have triumphed over both economic and housing policies in the area of tax relief, at the time of writing the desire for public expenditure cuts appears to have triumphed over housing policy in the field of supplementary benefit payments. In the Green Paper on the social security review, (2) the government is suggesting that it might remove, or reduce at least for an initial period, the entitlement of owners on supplementary benefit to receive mortgage interest payments. It is suggested that the period should be covered by a mixture of insurance protection and concessions by lenders to mortgagors. The argument in the Green Paper is that the payment of the total cost of interest payments creates a more favourable position for those who are unemployed than for those who are in low paid work. Part of the unstated reasoning may also be that home owners are currently more favourably treated than tenants in that their assets, in the form of equity in a house are disregarded, whereas £3,000 of liquid assets removes eligibility for supplementary benefit. An arrangement whereby supplementary benefit recipients were required to draw upon this equity in the first six months of entitlement would create greater equality of treatment. However, this would put the lenders in a difficult position because they would be asked to make a further advance just at the point when the original one looked insecure and they would be eroding the equity which is their best guarantee against mortgage losses. In addition there are many first time buyers who already have little or no equity upon which to draw. Furthermore, while tenants on supplementary benefit receive their total housing costs, home owners already receive only part; in particular they receive only a token payment for repairs, maintenance and insurance.

Table 2: Mortgagors Receiving Regular Weekly Payments of Supplementary Benefit in a Week in November/December: Great Britain.

Year	Number of Mortgagors	Year	Number of Mortgagors
	'000s		'000s
1967	75	1979	98
1971	90	1980	134
1976 (a)	122	1981	196
1977 (b)	124	1982	235
1978	105		

(a) Approximate figures
(b) Change in method of estimation

Source: Department of Health and Social Security (1982 and 1983) Social Security Statistics, HMSO, London Tables 34 and 58.

Therefore though there is logic in the government's arguments for consistency, with the low paid (by implication) with tenants on supplementary benefit, there is a real question as to whether bringing owners into line with these two groups is consistent with a policy of giving 'as many people as possible the opportunity to become home owners'. (3) A cut in interest payments to supplementary benefit recipients would save much less than the removal of just the higher rate tax relief (4) and it would create maximum havoc amongst the most vulnerable buyers. Interest payments from supplementary benefit have had the effect of cushioning the unemployed against the loss of their home and stabilising house prices. The payments have created greater stability of ownership and allowed lenders to give mortgages to applicants who might otherwise have appeared a poor risk. The impact of these payments can be gauged if we look at the default situation during the depression, when there were no such payments. At the time home ownership was largely confined to the middle class and the elite of skilled manual workers, and yet in 1933 the Co-operative Permanent Building Society (now Nationwide) found itself possessing 2.35 per cent of all its outstanding mortgages (table 3). With present levels of home ownership such levels of mortgage possession would mean that 150,000 building society borrowers would have been possessed in 1984, as compared with an actual total of 11,000. Not all this difference can safely be ascribed to supplementary benefit interest payments but clearly the payments have been a major stabilising factor. Whether or not the proposed changes are implemented will depend very much on the degree to which government recognises the damage it would cause to its home ownership policy and to its relationships with the building societies. The government's main argument for change, namely the mortgage interest payments to the unemployed act as a 'discouragement for owner occupiers to return to work' (5) is a clear illustration of the way in which the philosophy of monetarist economic and social policy conflicts with a housing policy based on high levels of home ownership, for which adequate levels of income maintenance are essential to provide a climate of confidence and stability for lenders and buyers. Furthermore, there is no evidence to substantiate the government's belief that mortgage interest payments deter the unemployed from seeking work.

The biggest contradiction in the government's approach to housing policy must be its attempt to in-

Table 3 : Mortgage Possession in the Depression: Co-operative Permanent Building Society

Year	Total Balance Outstanding	Total Loans	Average Outstanding Mortgage	Proceeds of Sale of Properties in Possession	Estimated No of Properties Possessed	Proceeds of Sales of Possessed Property as % of Outstanding Balances
	£		£	£		
1928	6,168,218	14,054	439	23,879	54	0.39
1929	8,536,974	18,713	456	34,265	75	0.40
1930	11,306,607	23,832	474	72,328	153	0.64
1931	13,644,023	28,356	481	168,489	350	1.23
1932	14,571,217	30,744	474	309,220	652	2.12
1933	15,776,970	33,838	466	370,665	795	2.35

(Statistics discontinued)

Source: Co-operative Permanent Building Society Annual Accounts kindly supplied by Nationwide Building Society.

crease home ownership rapidly and profoundly at a time when its economic policy accepts, if not creates, large-scale unemployment. By its very nature, home ownership flourishes best in periods of high, stable employment. Moving from one owner occupied home to another is slow and costly and not compatible with an economic policy of 'getting on your bike'. Stable home ownership requires an expectation that house prices will not fall; the risk of large-scale local redundancies and high interest rates undermines such an expectation. Most of all, a drive to increase lower income home ownership is undermined if an increasing number of those who buy subsequently lose their homes as a result of mortgage default.

Despite the cushioning effect of social security payments rapidly increasing numbers of buyers have been losing their homes. As more and more people have become not just unemployed, but long term unemployed, so the level of default has grown. In January 1985, 34.9 per cent of all the registered unemployed in the United Kingdom had been unemployed for more than a year. Of those aged 25 to 54, the proportion was 44.9 per cent and for those aged 55 and over 52.5 per cent. (6) With government's decision not to allow the unemployed to have the long-term rate of supplementary benefit, the ability of the unemployed to stay solvent has been further undermined. (7) Research has shown that the situation of families with children on supplementary benefit has been particularly difficult because of inadequate scale rates. (8) Again measures aimed at increasing the incentive to work and reducing public expenditure have had their impact on home ownership policy.

The number and proportion of mortgages taken into possession by building societies have quadrupled between 1979 and 1984. So have the number and proportion of mortgages which are over 6 months in arrears. In 1984, 10,950 mortgagors had their homes taken into possession compared with 2,530 in 1979 (table 4). In 1984, there were 50,150 mortgagors (7.9 in every thousand) who were more than six months in arrears compared with about 10,950 in 1979. In 1984 54,754 owners were taken to the county court for possession, and 35,397 had possession orders made against them. (9) Then there is the unknown, but undoubtedly large number of people who sell their homes when they get into arrears, rather than wait for court action. Large proportions of the people who lose their homes in any of these ways have to turn to the local authorities for help, just at a time when local authorilties' resources are being severely cut. In some parts of the country mortgage possession is now the commonest cause of homelessness.

One of the reasons for the high rate of possessions and particularly the high rate of possessions by the bailiff (7,963 cases in 1984) (10) is that houses in many areas have become difficult to sell at all, or quickly enough to clear a debt or for a price that will cover the outstanding debt. Thus the ability of an owner to sell in advance of possession is eroded. Sluggish house prices and low rates of inflation have reduced the equity that owners can expect to have in their property after a few years of ownership. Ironically while the Conservative government has as its prime economic target the reduction of inflation, high rates of inflation were precisely one of the factors, along with low real interest rates, that fuelled the growth of home ownership in the 1970s and made it such a 'good buy'. Home ownership in the 1980s with high real interest rates and low inflation is quite a different story. The public's assumption that home ownership equates with security of tenure is being undermined by a combination of changed economic circumstances and the expansion of home ownership to households with lower and less stable incomes. This is bound to have some effect on public attitudes towards ownership.

The other assumptions about home ownership, which have been particularly important to government are also being eroded. The first is crucial to employment, namely that home ownership facilit greater labour mobility than does council renting. This assumption has been important because it made increased home ownership seem highly compatible with employment and social security policies with increasingly stressed work incentives for the unemployed. The second is vital to housing policy and to public expenditure considerations, namely that standards of repair by home owners are much better than by landlords and so the stock is maintained at lower cost to the Exchequer.

It was the miners' strike that brought home to the government the fallacy about labour mobility. Instead of home ownership deterring the miners from striking as conventional wisdom suggested it should, the thought of having to sell their homes as well as move job gave the miners a greater stake in trying to retain jobs in their communities. In a choice between labour mobility and work incentives and home ownership, government has so far apparently leant towards labour mobility, and work incentives since these are the rational for its proposal to cut mort-

Table 4 : Summary, Arrears and Possessions, All Building Societies

Period	No. of Loans at End of Period	Properties Taken into Possession in Period		Loans 6-12 Months in Arrear, End Period		Loans over 12 Months in Arrear End Period	
		No	%	No	%	No	%
1979	5,264,000	2,500	0.048	8,420	0.16		
1980	5,396,000	3,020	0.056	13,490	0.25		
1981	5,505,000	4,240	0.077	18,720	0.34		
1982	5,664,000	5,950	0.105	23,790	0.42	4,810	0.085
1983	6,018,000	7,400	0.123	25,880	0.43	6,620	0.11
1984	6,348,000	10,950	0.172	41,900	0.66	8,250	0.13

Notes. 1. The figures are based on statistics provided by some or all of the 17 largest societies (16 from the end of 1982) which accounted for 84.2 per cent of all outstanding mortgages at the end of 1983. The figures have been grossed up to represent the whole industry by reference to the number of outstanding mortgage loans as published by the Chief Registrar of Friendly Societies. The figures for 1983 and 1984 are BSA estimates. (It should be noted that the figures refer to the number of mortgage loans and not to the borrowers).

2. The figures for possessions are based on a sample of 11 societies up to the end of 1981 and there is therefore a slight discontinuity in the series at this time.

3. Properties taken into possession include those voluntarily surrendered.

4. The figures for loans 6-12 months in arrear prior to the end of 1982 should be treated with considerable caution.

Source: Building Societies Assocition Mortgage Repayment Difficulties, Report of the Working Party on (Chair M. Boleat) BSA, London 1985.

gage interest payments to those in supplementary benefit. (11) However in reality the proposal looks more like a simple public expenditure cut.

More fundamentally home ownership is currently a problem for labour mobility because of widespread reliance on two earnings to meet mortgage payments. The civil service and particularly departments such as the Ministry of Agriculture which have many regional 'out stations' are finding it increasingly hard to operate promotion schemes which rely on people moving from one area to another. Civil servants are frequently turning down offers of promotion which involve a move, especially to higher priced areas, notably London and the South and South East. Prices in these regions have been rising noticeably more rapidly than elsewhere. (12) Those within these markets dare not come out, even temporarily, and those outside them cannot afford to enter. But even without price gradients between regions there is a particular problem, namely, that many mortgagors rely on two sets of earnings and if the wife cannot also be guaranteed a job, the move would involve at the very least a substantial drop in housing standard or financial crisis. The problem for the middle classes is possibly greater because of the higher grade and income, of wives' jobs, and therefore their relative scarcity, but the problem is a general one.

For lower income buyers one of the biggest problems of mobility is the length and cost of the buying and selling process. Unlike professional and management jobs where it is assumed that there will be a delay between the appointment and the appointee taking up the post, most manual or lower level non-manual jobs involve an immediate start. If the job is in another area this may mean keeping two homes running for at least three or four months. In extreme cases some building societies have been finding that buyers with very little equity in their property just move away and return the keys, leaving it to the society to sell the house and deduct costs and outstanding payments.

Arguably a government interested in labour mobility might be expected to give much greater stress on the revival of the private rented sector. However, despite government's strong rhetoric about reviving this sector, in reality lip service only is paid by this government to measures to make private renting profitable. While landlords are treated so much worse than owner occupiers for tax purposes and while fair rents exclude consideration of return on capital, no revival of private renting can be expected. Merely allowing rents to rise would not however, solve the incentive problem. Essentially the greater the financial inducements to owner occupation and the further down the income scale owner occupation is brought, the less potential tenants there will be who can afford rents that would give a reasonable return. Few private landlords are interested in incurring the management costs involved in housing tenants who have difficulty in paying the rent.

The labour mobility problem is very real and presents dilemma for government but it is not visible and therefore constitutes little in the way of a political crisis. The issue of repairs and maintenance in the owner occupied sector is much more visible, and constitutes a crisis for government, both by undermining the ideology of home ownership and by making severe calls on public expenditure. In essence the problem is that very many owners are unable to afford routine repairs and maintenance on their homes. Crisis point is reached where the properties owned are also in need of modernisation and improvement. Two groups are particularly involved, the elderly and owners of pre-1919, often inner city, houses. In some areas the two categories substantially coincide, in others not. For instance, in Birmingham's inner city owners are increasingly younger, low income Asian households, (13) whereas in South Wales the valleys are no 'inner city' but they have large concentrations of elderly owners in very deteriorated terraced housing. Quite apart from the pre-1919 stock there are massive repair problems for elderly people in interwar estates and in retirement resorts.

According to the 1981 English House Condition Survey (14) 49 per cent of all unfit dwellings in serious disrepair were occupied by elderly people (table 5) and 62 per cent by people outside the labour forces (table 6). But even those in full-time work have been finding repairs and maintenance an increasing strain. In the period 1970 to 1980 the cost of repair and maintenance rose 390 per cent while the median earnings of full-time male manual workers rose only 308 per cent. Average expenditure on repair and maintenance by owner occupiers rose only 291 per cent (table 7). By 1981 the average cost of outstanding repairs required per owner occupied house was £1,883 and the total national repair bill in the owner occupied sector was £19.4 billion. (15)

Deteriorating house condition is probably the issue over which the contradictions of government policy are most conspicuously revealed. Ministers

Table 5: Household Type by Condition

	Satisfactory Dwellings	Unsatisfactory Dwellings				
		Fit			Unfit	
		All Amenities Medium Disrepair	Lacking Amenities Low/ Medium Disrepair	Serious Disrepair	Low/ Medium Disrepair	Serious Disrepair
Single Person under 60 years	5	7	8	8	11	10
Small adult household	13	12	6	12	11	7
Small family	21	18	14	15	12	10
Large family	7	8	5	8	7	8
Large adult household	22	23	12	22	16	16
Older small Household	18	18	27	19	19	22
Single Person under 60 years	14	14	28	16	24	27
All households	100	100	100	100	100	100

Source: Department of the Environment (1981) <u>1981 English House Condition Survey</u>, Part II, Table 3.

Table 6: Working Status of Head of Household by Condition

	Satisfactory Dwellings	Unsatisfactory Dwellings				
		Fit			Unfit	
		All Amenities Medium Disrepair	Lacking Amenities Low/ Medium Disrepair	Serious Disrepair	Low/ Medium Disrepair	Serious Disrepair
Full-time Empl.	62	56	34	53	46	38
Part-time Empl.	4	4	3	5	4	2
Looking for Work	4	5	3	4	6	6
Retired	22	25	45	27	30	38
Housewife/Other	8	10	15	11	14	18
All Households	100	100	100	100	100	
Median Income Head and Partner	£5566	£3937	£2238	£4677	£3077	£2234

Source: Department of Environment (1981)
1981 English House Condition Survey Part II, Tables 8 and 9.

Table 7: Repairs and Maintenance: Expenditure and Costs Indices: Great Britain (1970 = 100)

	Average Expenditure on Repairs and Maintenance by Owner (1)	Cost of Housing Repairs and Maintenance (2)	Repairs and Maintenance Activity Ratio	Retail Prices (2)	Durable Household Goods Prices (3)	Median Earnings Full Time Male Manual Workers (4)
1970	100	100	1.00	100	100	100
1971	116	108	1.07	109	107	110
1972	110	116	0.95	117	112	122
1973	140	147	0.95	128	118	142
1974	184	179	1.03	148	135	163
1975	184	222	0.83	184	165	298
1976	252	261	0.97	215	181	243
1977	240	296	0.81	249	209	265
1978	258	327	0.79	270	229	299
1979	357	405	0.88	306	253	343
1980	391	490	0.80	361	284	408

Source: (1) Family Expenditure Surveys

(2) Department of Environment Supplementary Memorandum to House of Commons Environment Committee Table 69

(3) Annual Abstract of Statistics; Morley, S. (1977) Housing Supply, PCL.

(4) Office of Population Censuses and Surveys; Social Trends.

\+ Source: Doling, J. Owner Occupation, House Condition of Government Subsidies, in Booth, P.A. and Crook, A.D.H. (eds.) "Low Cost Home Ownership: An Evaluation of Housing Policy Under the Conservatives" Gower, Aldershot, 1985.

state that the whole point about home ownership is that people must take responsibility for their own homes, yet that position, taken to its logical conclusion, would involve not just continued massive deterioration of the stock but also a growing reluctance of building societies to lend in older areas and even the compulsory purchase of owner occupied property. It was in fact the prospect of having to compulsorily purchase hundreds of owner occupiers' homes, in order to achieve the improvement targets set in housing action areas, that contributed to the decision of the Conservative controlled council in Birmingham to initiate its enveloping scheme.

So electoral issues intervene. It would, needless to say, be electorally risky for a local authority to be compulsorily purchasing home owners. In contrast giving out improvement and repair grants is electorally advantageous. So in the run up to the 1983 election repairs grants were liberally available, enveloping received government blessing, and there was record government spending on private sector improvement and repair. Since then public expenditure considerations have become paramount. Enveloping has been reduced to a pale shadow of its former self, improvement and repair grants have been slashed and new proposals have been put forward to the Government's Green Paper on repair and improvement to target grants very tightly on the poorest and to use loans as a substitute. (16) Effectively these proposals ignore the very profound problems faced by the owner of a deteriorated house, when that owner's income is anywhere from supplementary benefit level to somewhere around the average industrial wage. Such owners seldom have the resources to improve their homes without financial help.

Rising Costs of Home Purchase

Conservative policy on home ownership has not attempted to tackle the problems of mortgage and maintenance costs after purchase. But what government has stressed are measures to make it possible for more people to buy. The main plank of this policy has been council house sales; 196,680 properties were sold in 1982 and 133,820 in 1983. (17) But there have been a number of other small subsidised schemes, particularly build-for-sale, improve-for-sale and home-steading. These schemes including council house sales have all used subsidies to reduce the purchase price. ownership has been different in that it reduced monthly payments but not the purchase price.

However, all these initiatives except council house sales, have been very small compared with the volume of houses being built for sale in the private sector in 1982 and in 1983. (18) So the success of the policy of making entry costs lower has to be seen in relation to main market conditions not the special case of government schemes. Here again there are problems, for prevailing high interest rates and the pressure on housing and land in areas where employment is available have produced high and rising buying costs.

In 1970 the average price of a new house in Britain was £5,180. (table) By 1980 it was £27,244 and by 1984 £35,128. If we take median gross weekly earnings of full-time male employees, in 1970 the price of a new house was 3.7 times the median; in 1980 it was 4.7 times the median and in 1984 it was 4.2 times the median (see table).

However, price is not the only variable in the cost of property. The others are the mortgage interest rate and mortgage term. A mortgage of £10,000 over 25 years at 13.00 per cent, the recommended rate of interest in Britain in 1985 cost £113.70 a month or £1,364.40 a year. At the rate prevailing in Britain in 1970, 8.50 per cent the same loan cost £80.15 a month or £961.80 a year, or £402 a year less in money terms. The prevailing interest rates are therefore as an element of the cost of purchase. To control the supply of money and more recently to stabilise the sterling exchange rate, nominal interest rates have been raised. As we will also see later, mortgage interest rates have increased in line with other interest rates because of the competition between lenders for investors More crucial are <u>real</u> mortgage interest rates, namely the difference between the rate of inflation and the mortgage interest paid. (figure 1.0) These have been rising since the mid-seventies to a record level in 1984 and 1985 of over 6 per cent and are making the issue of 'affordability' in the owner occupied sector one of growing concern to government, lenders and buyers alike.

For first time buyers in particular there has been an adverse trend in the relationship between their incomes and the cost of the average newly constructed property. A combination of high prices and high interest rates has shifted the ratio of mortgage payments (before tax relief) to incomes of first time

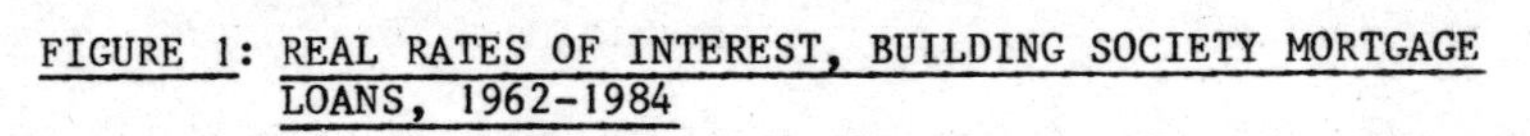

FIGURE 1: REAL RATES OF INTEREST, BUILDING SOCIETY MORTGAGE LOANS, 1962-1984

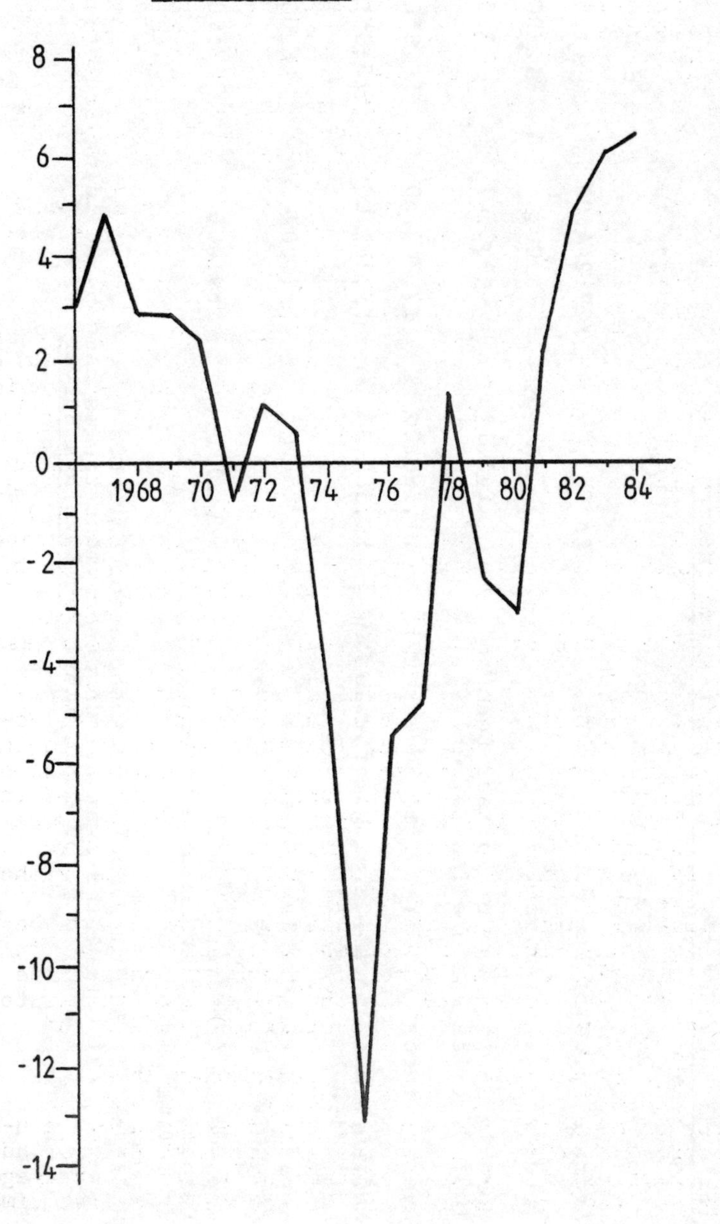

Table 8: House Prices, Interest Rates and Incomes

	1970	1978	1984 3rd Quarter
Average Price of New House (1)	£5,180.00	£27,244.00	£35,128.00
Interest Rate at Time (1)	8.50%	14.00%	12.00%
Monthly Repayments for 25 Year Loan on 100% of Price (Before Tax Relief) (2)	£42.17	£330.47	£373.41
Average Income of First Time Buyers (1)	£1,766.00	£7,749.00	£9,850.00
Median Gross Annual Earnings of Male Full-Time Employees (3)	£1,388.00	£5,762.00	£8,351.00
Repayments (Before Tax Relief) as Percentage of Income of First Time Buyers	29.00%	51.00%	46.00%
Repayments (Before Tax Relief) as Percentage of Median Earnings of Male Full-Time Employees	37.00%	69.00%	53.70%

Sources: (1) Building Societies Association Bulletin
(2) Calculated from standard repayment tables used by Building Societies.
(3) Department of Employment, New Earnings Survey.

buyers from 29 per cent in 1970 to 51 per cent in 1980 and 46 per cent in 1984. (table 8) This problem has produced, first, a lowering of standards in housing newly constructed for first time buyers and, second, a tendency for builders to leave the first time buyer market and move into construction for those moving up market. In 1970 first time buyers bought houses which were on average 74 per-cent of the price of houses bought by previous owners. By 1980 this ratio dropped to 61 per cent and remained at that level in 1984. (table 9)

The problem of budget stretch for even middle income buyers, particularly in high priced regions of the country makes the reform of tax relief difficult to tackle. This is because any new income related subsidy for owners which was roughly equivalent to even a more generously defined housing benefit in the rental sector, would be unlikely to be able to come far enough up the income scale to compensate middle income owners for the loss of their relief.

Table 9: Price of First Time Buyers Homes as Percentage of Price of Houses Bought by Former Owner-Occupiers

1970	74.2	1977	66.8
1971	72.6	1978	63.9
1972	67.9	1979	62.0
1973	66.5	1980	60.5
1974	69.3	1981	60.3
1975	69.1	1982	58.0
1976	67.2	1983	56.9
		1984	61.0

Source: Calculated from Building Societies Association Bulletins

The Building Societies

In recent years building societies have appeared to be willing instruments of government housing policy. Indeed they have contributed to the support of the current emphasis on home ownership as the major focus of policy by conducting their own research demonstrating the strength of the desire for home ownership. (19) As individual institutions many of them have been eager to expand their activities. They have sought to increase their share of the savings market, and then subsequently increased their lending and with it contributed to the expansion of owner occupation. But many societies have also been eager to enter into activities, outside the traditional mortgages to individual house buyers, such as assurred tenancy schemes and urban renewal.

However, the costs of this role are now becoming more evident. As government is unwilling to recognise or act upon the weaknesses in its policies, so the building societies have found themselves having to draw upon their own resources to act as a buffer between financially stressed owners and unhelpful government policies.

This presents building societies with a problem. Despite their recent higher profile on housing issues, their primary allegiance remains to their investors. These investors can only be attracted by a good and safe return on their money. This requires a commercial approach to lending, arrears management etc.

In recent years the building societies have shifted towards more competitive interest rates for investors. Previously the agreement over interest rates led to lower and more stable interest rates to borrowers but unattractive rates to investors and hence to mortgage queues; in effect investors were subsidising borrowers in the late 1960s and 1970s. The decision to shift to competitive rates to investors has ended mortgage queues but it has added to the cost of a mortgage. (20) The more people enter home ownership rather than renting, the more mortgage money will have to be raised and the more the building societies will have to seek out investors to provide that money. The fact that there are many building societies competing amongst themselves is also likely to raise the rate because it is easier to switch societies as an existing investor than as an existing borrower.

At the same time, it is important to building societies to retain their image as caring lenders. In the 1960s and 1970s building societies were heavily criticised for the restrictive nature of their lending, in particular their neglect of inner cities, black buyers, women and single people. (21) In the 1980s, though the criticism about the poorest parts of inner cities and about black buyers are still made, many building societies have noticeably relaxed their lending criteria on older property, lower income buyers, joint owners, women buyers and single people. This relaxation has come not just in response to the criticisms mentioned and to sex and racial discrimination legislation, but more importantly as an adjustment to the growing scarcity of rental alternatives, to the ending of mortgage queues and to competition for borrowers as well as investors.

So building societies, in reaction to their own internal requirements and those of the housing market, and in support of government home ownership policies (not just those since 1979) have departed substantially from the extreme caution which characterised their lending in the mid 1970s. It was inevitable that this shift in the style of lending would produce marginally greater risks of arrears. However, in the economic conditions of the late 1970s, with rising house prices, high rates of inflation and, by present standards, relatively low rates of unemployment, the risk of actual loss to societies was very small indeed. In most cases the owner's equity in the property was sufficient to cover the debt. However, more liberal approaches to lending have begun to take on more significance as unemployment has grown and as inflation rates have fallen. In this new economic climate, probably the most contraversial features of the relaxation of lending criteria have been income-to-loan ratios and percentage loans, particularly the latter. As house prices have risen in relation to incomes, and as lending to younger buyers has extended so building societies have been giving larger percentage loans (figure 2). The 90 per cent loan to first-time buyers has become commonplace and 100 per cent loans are not the rarity they used to be. But as arrears and possession grow so building societies are becoming more aware of the default problems associated with buyers with high percentage loans. Not only are such buyers more likely to incur arrears but, more important, they are also more likely to build up a debt greater than the sale value of the house.

By 1984, arrears, as we have shown, had reached a level which was worrying to building societies. (22) So much so that they set up a working party on the subject. There were still relatively small losses on possessions (about £2.5 million just on these accounts with 6 to 12 months arrears (23)) and the cost of arrears management was estimated at £15 million a year. (24) The greatest risk of arrears was found to be in the first few years of the mortgage, which

FIGURE 1.2: AVERAGE ADVANCES, BUILDING SOCIETIES, 1969-1983

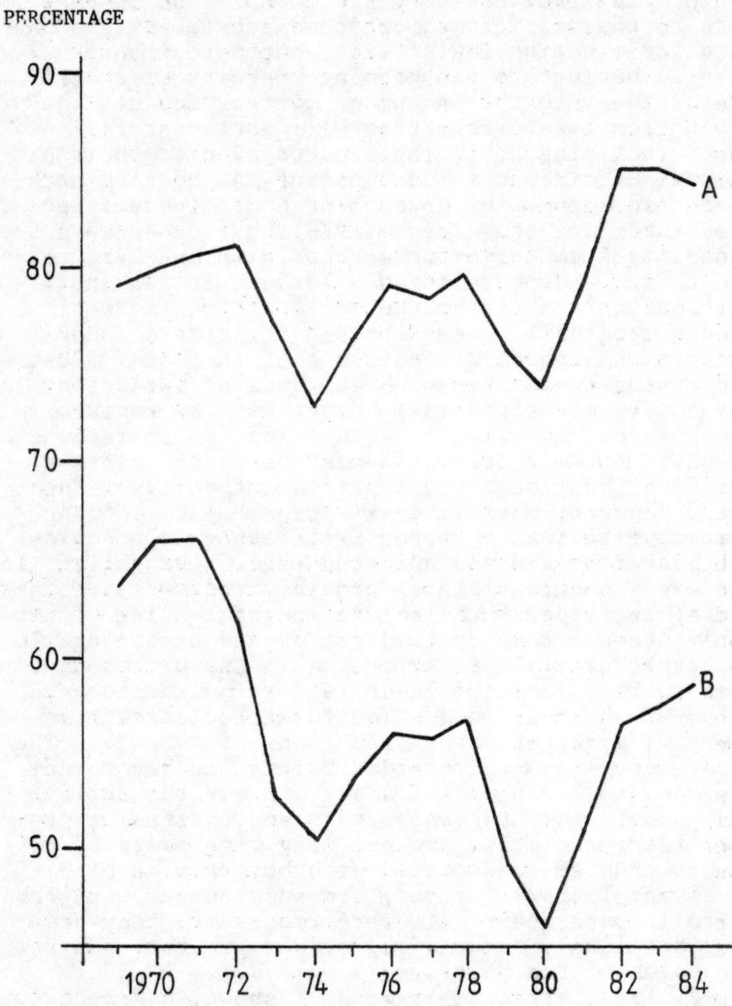

KEY

A = Average percentage advance - first-time buyers.

B = Average percentage advance - former owner-occupiers.

Source: Building Societies Association Bulletin, 1984

suggested that more stringent income and percentage loan criteria might be the solution.

There have been several inbuilt safeguards which have allowed building societies to continue to pursue more risky lending practices than might otherwise have been the case. First and foremost, has been the provision in supplementary benefit for the whole of the interest on a mortgage to be met, in effect indefinitely. This provision has become increasingly important to the building societies in cushioning the impact of unemployment. But it can also be seen as having led building societies into types of lending that they might otherwise have avoided, notably large percentage loans.

The same can be said of improvement and repair grants. These have been crucial in encouraging building societies to lend on older property in inner city areas. Without the prospect of grants, lenders would have been too anxious about the maintenance of the value of the property in which they were investing.

The proposals to remove or reduce interest payments under supplementary benefit and to replace improvement grants with loans except for the poorest owners have therefore been a considerable shock to the building societies. As the General Manager of Nationwide commented (25) a partnership between the building societies and the government requires effort on both sides. But, if building societies felt let down on the Green Paper proposals, the government in its turn, felt that the building societies had helped to prolong the miners' strike by not enforcing the payment of arrears.

Another source of complaint both by the building societies and by the CBI (26) has been the continuing high level of interest rates produced by government economic policy. These very high rates contribute to the default problems of buyers.

So building societies and government are both in the position that they would like a reduction in the costs of buying but other priorities weigh more heavily, in the building societies' case growth of investment and in the government's case other economic and social priorities. Each tends to regard it as the other's responsibility to keep the cost of buying down.

All these problems present building societies with a dilemma. Should they be more cautious with their lending, and incur criticism for not supporting government home ownership policy and for denying people homes in areas where there is little rental housing. Or should they, as now, lend relatively easily and risk rising default and possession. Too high a rate of possession could tarnish their image not only with buyers but with investors. Ironically, the safe-as-houses image of home ownership now depends heavily on the fact that hard pressed authorities house so many homeless ex-home owners. The fact that so many thousands are absorbed quietly into the public sector makes them less visible as a blot on the home ownership record.

Conclusion

We can then sum up as follows:

1. The government finds itself committed to the oversimplified statements of its own rhetoric on home ownership, yet the economic climate is increasingly exposing the fallacies behind the rhetoric.
2. Its economic policies exacerbate rather than diminish the problems of home owners.
3. One logical route out of that contradictory position, namely the reform of housing finance is blocked by short-term electoral considerations.
4. Failure to recognise the need to adjust housing policies to the economic climate will increasingly undermine both the 'safe' image of home ownership and government's policies to promote it.
5. The tensions created by this conflict between policy and economic realities are likely to affect adversely the relationship between government and the main agents of home ownership policy, the building societies.

The people most crucially caught up in the contradictions of government economic policy and housing are of course the buyers. The essential contradiction for them is that just when there is a government whose policy is addressed towards giving home ownership to as many people as possible, home ownership is losing some of the features that made it so attractive. Or rather it is being revealed as never having had those attractive features when available to the people towards whom it is now being extended, namely lower to moderate income households. In surveys about tenure preferences the features which people usually single out as the most important attribute of home ownership are its investment value, followed by independence and security in a 'home of your own', and the quality and type of property available, as compared with that in the public sector. Essentially these have all been characteristics associated with higher priced, higher quality suburban housing. When ownership also covers lower income buyers, particularly of inner city terraces, the stereotypes no longer apply so clearly. For these buyers the quality of housing may be worse than council housing, the investment value less certain than in suburban areas and more vulnerable to changes in government policy on improvement grants and the tenure only as good as the owner's employment stability.

So as home ownership being more widely available, it is likely that its image will change to a less idealised one and those who are on the borderline of buying will more realistically weigh up the risks of owning against benefits. In this way, the very success of spreading home ownership could sow the seeds of a reaction against it and a revival, albeit modest, in the popularity of renting.

NOTES AND REFERENCES

(1) Hansard, Written Answers, cols. 505-8, 20 March 1985.

(2) Secretary of State for Social Services, Reform of Social Security: Programme for Change vol. 2 pp. 25-6. Cmnd. 9518 HMSO, 1985.

(3) Op. Cit. p. 26.

(4) Removing the higher rate of tax relief would 'save', that is increase tax revenue to the Exchequer, by about £160 million per annum at 1984/85 levels of income and mortgage interest. Hansard, Written Answers, col. 400, 12 June 1984.

(5) Secretary of State for Social Services loc. cit.

(6) Department of Employment Gazette vol. 93, no. 4, April 1985.

(7) Doling, J. Karn. V. and Stafford, B., 'Unemployment and Home Ownership' Housing Studies, vol. 1, no. 1, 1986.

(8) Policy Studies Institute, The Reform of Supplementary Benefit, Research Papers 84/85, 1 and 84/5.2, PSI, London 1984.

(9) Communication by the Lord Chancellors office The publication of these statistics in Judicial Statistics has been discontinued.

(10) See note (7)

(11) Secretary of State for Social Services, op.

cit.

(12) Building Societies Association, *Building Societies Association Bulletins*, BSA, London.

(13) Karn, V., Kemeny, J. and Williams, P. *Home Ownership in the Inner City: Salvation or Despair?* Gower, 1985.

(14) Department of the Environment, *The 1981 English House Condition Survey*, HMSO, 1981.

(15) Op. cit. estimated from part 1, table 21.

(16) Secretary of State for the Environment, *Home Improvement: A New Approach*, Cmnd. 9153, HMSO London, 1984.

(17) Department of the Environment, Scottish Development Department and Welsh Office, *Housing and Construction Statistics*, September quarter, part 2, no. 20, 1984.

(18) Ibid.

(19) Building Societies Association, *Housing Tenure*, BSA, London, 1983.

(20) Bank of England, 'The Housing Finance Market: Recent Growth in Perspectives' *Bank of England Quarterly Bulletin*, vol. 25, no. 1, March 1985, pp. 80-91.

(21) cf. Burney, E. *Housing on Trial: A Study of Immigrants and Local Government*, Institute of Race Relation, Oxford University Press, 1967.

Daniel, W. *Racial Discrimination in England*, Penguin Books, Hammondsworth, 1968.

Weir, S. 'Red Line Districts, *Roof*, July 1976, pp. 109-114.

Lambert, C. *Building Societies, Surveyors and the Older Areas of Birmingham*, Working Paper no. 38, Centre for Urban and Regional Studies, University of Birmingham, 1976.

Karn, V. 'The Financing of Owner Occupation and its Impact on Ethnic Minorities', *New Community*, vol. VI, nos. 1 and 2, Winter, 1977/78, pp. 49-63.

Karn, V. 'Low Income Owner Occupation in The Inner City', in Jones C., (ed.) *Urban Deprivation in Inner City*, Croom Helm, 1979.

(22) cf. Stevels, L. et al., *Race and Building Society Lending in Leeds*, Leeds Community Relations Council, Leeds, 1982, and Karn, V., Kemeny, J. and Williams, P. op. cit.

(23) Building Societies Association, *Mortgage Repayment Difficulties*, Report of the Working Party on (Chair, M. Boleat) BSA, London, 1985.

(24) Op. cit.

(25) Quoted in the *Guardian*, on July 3rd, 1985.

(26) E.g. account in the *Guardian* on July 8th, 1985.

Chapter Six

HOUSING AND CLASS IN THE INNER CITY

David Byrne

The first half of this chapter presents a review of current debates on the relationship between housing tenure and the basal determinants of the contemporary character of the class system. This is followed by a detailed case study of an area of inner city working class owner occupation in central Gateshead. The case study provides empirical material against which the argument reviewed in the first section may be set. In the conclusion I return to these arguments in the light of the empirical material presented in the case study.

I begin with two quotations:

> ...class is not this or that part of the machine, but the way the machine works once it is set in motion - not this interest and that interest, but the friction of interests - the movement itself, the heat and the thundering noise. Class is a social cultural formation (often finding institutional expression) which cannot be defined abstractly in isolation, but only in terms of relationships with other classes; and ultimately, the definition can only be made in the medium of time - that is, action and reaction, change and conflict. When we speak of a class we are thinking of a very loosely defined body of people who share the same categories of interests, social experiences, traditions and value systems, who have a disposition to behave as a class, to define themselves in their actions and in their consciousness in relation to other groups of people in class ways. But class itself is not a thing, it is a happening. (1)
>
> ...when we talk of 'the Base', we are talking of a process and not a state. And we cannot

> ascribe to that process certain fixed properties for subsequent translation to the variable processes of the superstructure. Most people who have wanted to make this ordinary proposition more reasonable have concentrated on refining the notion of superstructure. But I would say that each term has to be revalued in a particular direction. We have to revalue 'determination' towards the setting of limits and exertion of pressure, and away from a predicted, prefigured and controlled content. We have to revalue 'superstructure' towards a related range of cultural practices and away from reflected, reproduced or specifically dependent content. And, crucially, we have to revalue the 'base' away from the notion of a fixed economic or technological abstraction and towards the specific activities of men (sic) in real social and economic relationships, containing fundamental contradictions and variations and therefore always in a state of dynamic process. (2)

The above two quotations are the foundation stones of my contribution which is an attempt to say something about how housing and class are happening together rather than any sort of discussion of stratification. Let me straight away junk the term 'stratification'. It is far too static as a basic metaphor for the dynamic character of inequality in a social order. Tenure may still be useful and I will come back to that.

There is an extensive debate over 'class and tenure' which is essentially summarised in the paired articles by Saunders and Harloe in a recent issue of the IJURR. (3) One way to attempt this paper would be to evaluate that debate, to comment on it, draw from it, and see where it gets us. I am not going to do that. Instead I am going to start a synthesis more or less from fundamental elements, set the account generated against a specific experience, and then come back to the debate as it stands and see what it looks like in relation to this independent account. Let me emphasise that this is *not* a claim for originality. My way of looking at these things has been influenced by the terms of the argument *but* I have come to reject those terms as the basis of a coherent account and I think that the deliberate *fiction* of innocence is a useful *device* here.

The central concept I employ in attempting to understand what is going on, has gone on and will go on is that of the 'reproduction of labour power'. The emergence of industrial capitalism created an urban system, the first purpose of which was to provide means through which workers could purchase the commodities necessary to their continued capacity to work, from their wages. Among these necessaries was accommodation. A subsystem of capitalism very rapidly developed which was concerned with the production and realisation of this rather awkward thing - awkward because it involved land and was provided through an agency, housing, which was exceptionally expensive in relation to working class wages and in consequence brought in credit and hence the finance sector of capital. Very early in the development of this system the state acting as the 'collective' capitalist intervened in order to resolve the housing specific instance of the general contradiction within capitalism relating to the inadequate means generated for the reproduction of labour power, and especially the reproduction of <u>future</u> labour power. The actual complex mechanism which emerged from this is best described by the tenure label, private renting, although it must be stressed that what is meant is not:

> ...the legal recognition of the separate sets of social relations which intervene between the production of housing and its consumption as a use value (4)

but something more akin to what Stewart (5) is getting at in her explicit criticism of that formulation. Indeed I would go farther than Stewart who emphasises the production determined character of the legal form. What I would emphasise is that tenure has to be interpreted as a form of organisation of reproduction, rather than in any other way. Once it is recognised as such, we can, following the autonomist account of the role of the working class in <u>the</u> active contradiction within capitalism, understand the transformation of the tenure base of housing provision in the U.K. since the first world war in terms of being a consequence of class action.

Thus contemporary tenurial forms have to be understood as action generated, <u>but</u> these contemporary reproduction systems exist in relation to a productive base. Let me say straight away that I have a classificatory problem here. On the one hand I would tend to assign the reproductive processes contained within the operation of tenure to base as described above by Williams, leaving the cultural formulation of the significance of those processes of reproduction as content for superstructure. On the other I am about to argue that the nature of the processes of reproduction is determined, bounded and rather more than bounded, by the nature of relations at the point of production. Let me beg this question by suggesting that: production determines reproduction, and proceeding.

We need to get to 1984 quickly. I am here representing an account of the development of housing in all its aspects including tenure, in which central explanatory power is given to class action. This does not discount production relations in the production of housing as a commodity, but it sets them as subordinate in relation to the wider production relations in which, to use Friedman's terms, the working class on the whole, achieved centrality. (6) I should note that centrality as I am using it is extended beyond Friedman's formulation which was in specification production located. I am adding in an element of 'political centrality', of capacity to affect the course of events in the sphere of civil society, although this remains basally determined. The centrality of the working class in all its aspects depended on the virtual elimination of a reserve army of labour in the advanced industrial countries during the long boom and on the relation of labour and capital in a period of sustained growth.

That has now changed. In 1984 whatever we may mean by the word we would probably agree that we are in a crisis. I should note that I regard this crisis as itself a product of class action (see Byrne,(7) for an elaboration) but whatever its origins it is here. What a crisis is, is a turning point in which relationships are changed. The usual word is restructuring, which clearly has a lot in common with Mingione's (8) use of restratification', although I prefer the former because it gets us away from slabs and towards processes. The point can be expressed in context thus:

> The state of the regions economy could hardly be worse. Unemployment is higher than since the 1930s, and is the highest in Great Britain. The number of jobs for men continues to decline and last year the number of jobs for women also declined. There were job losses in all sectors of the economy - in primary, manufacturing and service industries. Worst of all the prospect is for continuing loss of jobs and increasing unemployment. (9)

In November 1984 there were 238,865 counted unemployed people in the northern region (18.7% of the working population). The real rate is probably closer to 25%. In Tyne Wear county there were 100,932 counted unemployed (19.9% of the working population). In September there were 17,967 counted unemployed in Gateshead MBC (18.4% working population). Tyne Wear County had 200,000 plus manufacturing jobs in the early 1960s. The figure is now less than 100,000 and falling. In Gateshead alone there have been 3000 redundancies in the first three quarters of this year. In Gateshead post-code areas unemployment rates vary from 30.5% to 8.9% although only 2 of the 25 post code areas are below 10%. I will be looking in some detail at the Saltwell-Shipcote area where the figure in September was 22.2%. (10)

The point is the economy is down the plug. What are the implications of this for the reproduction of labour power in housing? Let me discuss this in abstract first. For the capitalist mode of production housing systems (which can be labelled as tenures, although clearly tenures interact as components of a general system), are part of the process of reproduction of labour power, the terms of which are set at any point in time, and dynamically through time by the comparative strengths of classes in a war of position about the character and quality of reproduction. That is a gross simplification; first because it is not (often) a simple matter of two classes confronting each other, (although the complex of class fractions may be reduced to two classes, that is still a reduction); and second because part of the programme of one side has at times been to do with more than the nature of reproduction and has moved into a programme of transformation - quantity into quality. Nonetheless reproduction is central - reproduction, not consumption. If we use reproduction then we are constantly reminded that what is going on is to do with production because it is a process of preparing labour power.

This process is more than a beefing up of muscle, or even a skilling of mind. Despite its Althusserian provenance consider Cockburn's statement that:

> ...if capitalism is to survive each succeeding generation of workers must stay in an appropriate relationship to capital: the <u>relations</u> of production must be reproduced. Workers must not step outside the relation of the wage, the relation of property, the relation of employ-

ment. So 'reproducing capitalist relations' means reproducing the class system, ownership, above all reproducing a frame of mind. (11)

Remember Lefbvre's contention that the ideologies that matter are those that operate through practices, the rest are just games for the intelligentsia, (12) and consider Anderson's extended discussion of the modes of legitimation in the contemporary West and in particular the role of democracy itself. (13) How does tenure considered as a series of practices in reproduction operate now?

Let me put this directly. What if capitalist production does not need the labour power of workers on previous terms, but needs it on new, worse, terms; either actually or potentially? What about a situation in which previously central workers are being peripheralised - a situation in which emmiseration is a device in restructuring? What does this mean for the terms of which reproduction will occur? Can reproduction be unaffected? Does a change in the relation of labour and capital in production, a restructuring mediated in large part through space, have no implications for how housing is experienced? Let us go back to determination. When the 'relative autonomy of the political' was all the rage it was fashionable to talk about determination in the last instance. As E.P. Thompson points out there are a lot of historical instances when the lonely hour of the last instance actually arrives. In some institutional locations it is always present. No wage-labour system can operate an income maintenance system for the unemployed in which less eligibility however imposed, is not the central principle. The conditions of reproduction of the reserve army of labour through the wage substitute are determined narrowly. Housing is much more complicated but reproduction through housing cannot operate without the boundaries determined by production relations, and as production relations change so do the boundaries. There has been a reasonably developed discussion of this vis-a-vis public sector housing which looks at the marginalisation of the public sector. But what about owner occupation?

Ungerson and Karn (14) developed a very interesting discussion of the: 'apparently universal desire for owner occupation'. Let me paraphrase and selectively quote what seems to me to be a thorough and accurate inventory of reasons:

1. Houses have once more gained their ascendancy

as cast iron 'safe' investments - not this time for renting out, but rather for owner occupation. (15)

2. In response to fiscal crisis because...the IMF has yet to discover that tax relief is as important a subsidy as other visible transfers...Governments have been at pains to legitimise owner occupation and 'delegitimise renting of visible subsidised housing. (16)
3. Owner occupation has come to be seen not as relationship to property but simply as a higher standard of housing. (17)
4. Owner occupation, because it depends on stable repayment of loans by individual households over long periods, and because the market for houses for sale is dependent on the availability of loans for second-hand houses, literally purchases for the future. A future that 'works' is one that looks like the present, which is why owner occupation as a dynamic process is essentially conservative...the spread of owner occupation is a powerful means of purveying the values of the elite to other classes and hence of integrating classes into a shared value system...All citizens, it is suggested, have equal opportunity for such access to wealth - it is merely a question of seizing them. (18)

O.K. - all the above makes sense to me, so what happens in crisis? What are the effects on reproduction through owner occupation? It is time for a slice of life in this paper. I want now to look at, not a marginal, but a 'being marginalised' group of owner occupiers in Gateshead in order to pursue the implications of my argument that if we are to understand housing and class happening together then we had better work out the implication of developments in production for the nature and character of reproduction, and all this in the context of crisis which as O'Connor reminds us is a word from medicine describing that point in which the patient starts getting better or dies.

Shipcote and Saltwell

This area of central Gateshead consists of a mixture

of Tyneside flats (about 80%) and terraced houses built between 1890 and 1910. In all there are about 3,000 dwellings in this area although immediately adjacent parts of Bensham and Bede wards are essentially similar. This part of central Gateshead consists of the last major element of pre-1914 stock constructed for private renting to the working class in the district. It contains about two thirds of the 8,000 dwellings identified in the late 1970s in a programme of action for older dwellings.

Table 1: 1981 Census Index Comparisons

	Shipcote Ward	Gateshead	'Best' Council Estates	Suburban owner Occupied
% Pop. in Priv. H'holds 0-4	6.3	5.7	4.7	6.6
% Pop. in Priv. H'holds 5-15	14.3	16.0	18.5	18.5
% Pop. In Priv. H'holds Pens. Age	20.5	19.1	14.3	9.9
Fertility Ratio	26.8	23.9	19.8	23.5
Unemployed Males as a % Econ. Act	17.6	16.5	17.5	5.9
Unemployed Females as a % Econ. Act	7.0	7.3	7.7	4.4
% Pop. Econ. Act	57.9	59.6	63.0	69.0
% Pres. Perm. Priv. H'holds Owner Occupied	41.0	38.9	10.8	89.0
% Pres. Priv. H'holds L. Tenants	7.9	45.7	86.5	4.8
% Pres. Priv. H'holds PUF Tenants	42.0	7.9	1.2	2.5
% Pres. Priv. H'holds P. Fur. Tenants	7.3	3.5	0.2	1.0
% Pres. Priv. H'holds. H. Assoc. Tenants	1.3	3.0	6.7	0.4

Table 1 continued

	Shipcote Ward	Gateshead	'Best' Council Estates	Suburban Owner Occupied
% Pres. Priv. H'holds Business *Virtue of Emp.	0.5	0.9	0.6	1.3
% Pres. Priv. H'holds all Amenities Exclusive Use	90.8	97.8	99.1	98.7
% H'holds Density > 1.0	4.8	4.5	5.4	1.5
% H'holds only 1 Person	29.4	26.4	15.9	11.9
% H'holds with 6 or More Persons	3.7	3.3	5.1	1.8
% H'holds with no Car	66.0	58.8	59.7	21.6
% H'holds with 5 or More Rooms	35.7	48.4	68.0	74.6
% Pens. H'holds (ie Only Pens. Pres.)	28.4	26.2	18.6	13.0
% H'holds with Children	28.0	29.4	38.7	44.6
% H'holds Headed by Single Parents	4.5	5.8	6.6	2.7
% Adults Migrant in Last Year	4.1	10.3	8.6	7.8
% Tot. H'hold Spaces Purpose Built Flats	74.7	18.5	3.3	1.9
% H'holds Where Head is Econ. Act and in Social Class I	3.6	6.3	0.9	8.8
% H'holds where Head is Econ. Act and in Social Class II	12.5	21.9	8.8	33.0
% H'holds where Head is Econ. Act and in Social Class III (nonmanual)	22.1	14.4	10.3	17.8

Table 1 continued

	Shipcote Ward	Gateshead	'Best' Council Estates	Suburban Owner Occupied
% H'holds where Head is Econ. Act and in Social Class III (Manual)	40.5	33.8	45.9	31.1
% H'holds where Head is Econ. Act and in Social Class IV	16.3	15.5	24.2	7.4
% H'holds where Head is Econ. Act and in Social Class V	5.2	8.2	10.0	1.8

Source: 1981 Census Small Area Statistics

In Table 1 the area is compared with Gateshead M.B.C. as a whole; with the 'best council housing' Enumeration District set resulting from a cluster analysis of all enumeration districts in Gateshead where 60% or more of the households had council tenancies; and with a cluster of 'suburban owner occupied' enumeration districts resulting from a cluster analysis of all enumeration districts in Gateshead. From Table 1 it is immediately apparent that Shipcote is rather a 'type' area for Gateshead, being close to the Gateshead figures on almost all indices except housing tenure, housing forms dwelling size and social class of occupied heads of households. The housing related differences are a function of the distinctive housing form of the Tyneside flat in which what appears to be a single terraced house is in fact an upstairs and downstairs flat in a pair. Until very recently it was difficult to purchase a single flat, although this is now possible and a market has developed. In class terms, the area resembles the best council housing EDs although there are more nonmanual household heads. However, in general terms Shipcote is much closer to the 'good council housing' area profile than to that of the 'suburban owner occupied' cluster set. A final point worth noting from Table 1 is that dwelling size in the 'good council housing areas' appears to be comparable with that in the suburban owner occupied areas.

It is worth dwelling on the programme of action for older dwellings before proceeding. Probably the most important factor was not a function of housing policies as such but rather the impact of housing policy being used as a means to another end. I am referring to the period in the early 1970s when 75% grants were available in development areas, including Gateshead, for improvement work as part of an effort to reduce unemployment levels. Table 2 details the picture here both for Gateshead and nationally. In consequence of this end of slum clearance in the 1970s lack of amenities had become a residual problem by 1981 even in an area like Shipcote.

Formal declarations of General Improvement Areas and Housing Action Areas were much less significant. In the general area of Shipcote-Saltwell there were proposed two Housing Action Areas and two small scale clearance schemes. In addition a large area, the Avenues, was left with an indeterminate status. There was strong suspicion on the part of the residents that the then Leader of the Council and Chief Executive were looking to clearance but this was not supportable in terms of the condition of the dwellings. In practice the two clearance areas have not been proceeded with, and indeed many of the dwellings located within them have been improved using standard grants under the terms of the 1980 Housing Act. The GIA intentions have had little impact. There as been a programme of environmental improvements involving roads, pavements and fencing but this has been funded as part of the capital programme of the Inner City Partnership. The only real policy impact in the area is the consequence of the decision to exclude 1500 plus dwellings in the Avenues area from the serving of notices under section 9.1(a) of the 1980 Housing Act.

Section 9.1(a) Notices have in fact been the major method of intervention in areas of older housing in Gateshead in the recent past. Basically the serving of a section 9.1(a) notice is a procedure whereby an Environmental Health Officer serves a notice on the owner of a dwelling stating that while the dwelling is fit, it requires substantial structural repairs. The owner is then entitled to what was until March 1984 a mandatory grant of 90% of the cost of required works up to a ceiling of allowable cost of £4,800. Tyneside flats count as two dwellings and therefore the allowable cost is £9,600 per pair. As Table 2 shows repair grants, largely under section 9.1(a), have become the major grant element in Gateshead. As we shall see there has been a recent and

Table 2: Renovation Grants - Gateshead and England.

A) Gateshead M.B.C. (and Area Pre-1914)

	L.A. Tot.	Housing Assoc. Total	Priv.			
			Conversion and Improvement	Intermediate and Special	Repair	Total
1970						
1971						
1972						
1973						
1974	502	-	1,531	43	-	1,574
1975	334	-	670	1	-	676
1976	-	-	444	30	10	454
1977	781	19	564	6	17	585
1978	196	55	536	35	5	576
1979	312	31	430	26	11	467
1980	481	-	603	47	25	675
1981	178	2	344	175	42	561
1982	118	97	232	323	160	715
1983	287	261	350	369	455	1,174
1984 (1st $^3/_4$s)	383	13	209	155	1,201	1,565

B) England

	L.A. Tot.	Housing Assoc. Total	Conversion and Improvement	Intermediate and Special	Repair	Total
1970	40,357	2,472	26,097	45,196	-	71,293
1971	59,144	4,407	44,428	46,439	-	90,867
1972	97,482	3,406	78,524	45,634	-	124,176
1973	110,053	3,201	128,381	37,577	-	165,958
1974	73,494	3,952	164,525	27,823	-	192,348
1975	36,163	4,603	72,966	1,231	59	85,393
1976	38,983	13,388	57,784	10,849	85	68,718
1977	37,551	18,789	47,788	9,037	130	56,955
1978	60,871	13,056	49,424	7,935	211	57,758
1979	75,967	17,173	57,222	7,792	345	65,359
1980	77,275	14,832	65,809	8,143	513	74,465
1981	52,931	11,288	49,145	14,743	5,053	68,941
1982	57,222	17,286	54,732	20,600	28,696	104,028
1983	85,953	13,854	75,519	29,141	121,558	230,218

Sources: Housing and Construction Statistics and Local Housing Statistics.

dramatic change.

I want now to think about the way in which this area and the housing stock within it have functioned in relation to the system of owner occupation. My thoughts are based in part on an analysis of 1981 census data and local authority records, but principally on an opportunistic and totally unsystematic enthography which derives from having lived in the area from 1971 to 1983 and having been one of its local elected representatives since May 1984. In summary description the area is predominantly skilled working class, as it always has been; has both a high child population and a high proportion of pensioners; is overwhelmingly white and native Tynesiders despite the presence of a visible orthodox Jewish community and numbers of Asian small shop keepers; regarded as a stable and settled community; and has traditionally been the locale of the respectable working class of Gateshead. Traditionally the area voted Conservative, was a marginal seat in the postwar years, and last elected a Tory district councillor in 1979. The area seems to have begun to be transferred into the hands of owner occupiers post the 1957 Rent Act, although some of the terraced houses always had been in this tenure. (19) What had been gradual change became intensified in the late 1960s and the area's major function, at least in reputation, was as a locale where 'young couples' could buy relatively cheap terraced housing or pairs of flats, modernise with grants, sell at a profit and trade up. Indeed mere purchase of a modernised dwelling was often enough for trading up if you got in at the right time. We bought one of the larger terraced houses in this area in 1971 for £2,800, over the years spent about £3,000 at 1982 prices on structural repairs, and sold in 1982 for £16,000. Effectively we lived virtually rent free and made a year's net wages on the way. The terraced houses are much larger than anything available in the council stock and are well located in relation to amenities. Our operations in this system were at this time dependent on one relatively high wage which gave us a household income of about the same as our age peer households where the norm was for both husband and wife to work.

The massive extension of owner occupation in the 1970s was very dependent upon transition through places like Shipcote-Saltwell. My impression is that the people who did well were those of our age cohort - the immediately postwar born generation who entered the labour market in the full employment mid 1960s. This is supported by some preliminary findings of

Maura Banim based on survey and ethonographic work on an estate of the kind to which people moved on trading up. (20) There is no doubt that at this time owner occupation in this area offered both immediate and long term material advantages to those who took it up.

The immediate advantages were complex. Although the terraced houses are bigger than council houses in terms of floor space, the Tyneside flats are not and all lack gardens. In terms of amenity once improved there is little differences and pre-1914 stock is by repute easier to heat than semidetached houses. The point is that many of the postwar generation were moving from the Bevan houses in which their parents lived. It would be difficult to argue that their housing standards improved thereby. It was of course more complex than at first sight. These 'young couples' were establishing their own households. In the late 1960s they could not get access to the high quality stock in which their parents lived if they sought local authority tenancies. Instead they would end up in the 'Village', the appalling and about to be demolished system built maisonettes of St. Cuthbert's Village or similar locales. This was and is exactly what Dunleavy (21) meant by mass housing. As always we are dealing with constraints rather more than choices.

For those who moved through the system it all worked well. The particular area I am discussing was never 'red lined' by building societies, dwellings were easy to sell to others starting on the route up, and grants injected capital on a basis which was of course only very partially reflected in changes in the capital value of the dwellings themselves. In retrospect one is amazed at the instability of it all, even on its own terms. The capital gains were to such a very large degree dependent on negative real rates of interest and the whole shebang dependent on the constant arrival of first time buying 'young couples' with steady jobs which enabled them to get a mortgage, and on a high level of injection of capital by the state to improve amenity.

What is it like now? Well the problem is not one of lack of amenity but of a need for major structural repairs. These dwellings are getting on for one hundred years old. They almost all have slated roofs which are severely nail sick and in general they require large expenditure to restore their condition. They are in fact almost all proper subjects for notices under section 9.1(a). Until the 31st March 1984 this meant that most (but not all, and the exception was the product of the local authority policy cited above), householders were eligible for substantial grant aid and large amounts have been spent.

This has now been turned off like a tap. The local authority spent £7 million on private sector renovation grants in 1984-85, proposes to spend £5.4 million in 1985-86 and was committed to £1.5 million in each of the two following financial years. However this committed future expenditure is now problematic. The change in policy with regard to the treatment of local authority capital receipts meant that Gateshead had to cut £7 million from its 1985-86 capital programme and a major casualty has been the possibility of any new grants provision. Effectively a fairly arbitrary minority of residents have received assistance with structural repairs and most have got nothing. A survey by the environmental health department of the local authority has estimated that two thirds of these dwellings require major structural repairs, but are not unfit. About one fifth are getting them. The state is effectively withdrawing from grant aid to householders in this situation, despite assertions that grant aid is to be channelled to those most in need. At the same time mass unemployment has very rapidly dried up the stream of young couples who are forming new households through owner occupation. The 'on the way to clearnace' status of the Avenues area has just been reversed but it is most unlikely that adequate funds will be available for grant aiding. It is generally agreed that enveloping offers an excellent technical solution but the area is ineligible for HAA status on DOE guidelines because the level of amenities is too high so enveloping is ruled out.

In fact Ann Stewart's comment at the end of her discussion of Saltley identifies the housing issue here: "Those in need of substantial improvement will presumably be left to rot." (22) There are a lot of similarities between Saltely and Saltwell/Shipcote but there are some interesting and important differences which highlight the point about the process of state aid withdrawal being a process of marginalisation. The Asians of Saltley have always been marginal in the UK economy, or rather I should say peripheral, because the last thing they have been is unimportant. Their peripherality was and is a function of their historic role as migrant workers. In an important sense they were here to be marginal. My interpretation of Ann Stewart's account and of work by the Saltley CDP was that owner occupation there was a dead end.

For a lot of people (clearly by no means all) this was not true of Shipcote/Saltwell but is is true now. Not only are they 'trapped' by the collapse of trading up but many of them belong to precisely that section of the skilled working class who are being peripheralised by deindustrialisation expressed through redundancy and are thereby trapped in all senses. There have been very few repossessions for default in this area but the incidence of supplementary benefit paid mortgage interest is, so rumour has it (speculation, definitely), high. Certainly I have a fair amount of case work related to default on local authority mortgages. In many ways this is much more serious in relation to cultural politics, to the superstructural expression of the meaning of the tenure, than the Saltley situation. Owner occupation is turning round for these people. A process is going on in which the material and ideological (crude) conditions under which they are being reproduced is changing consequent upon their changed position vis-a-vis production. They are being marginalised here as well. This is where 'social being determines consciousness' comes in, but let me develop this in relation to the present debate.

Tenure and Class

Sanders had recently asserted that:

> ...housing tenure, as one expression of the division between privatised and collectivised areas of consumption is analytically distinct from the question of class; it is neither the basis of class formations - nor the expression of them - but rather the single most pertinent factor in the determination of consumption sector cleavages. Because such cleavages are in principle no less important than class divisions in understanding contemporary social stratifications, and because housing plays such a key role in affecting life chances in expressing social identity and (by virtue of the capital gains accruing to owner occupiers) in modifying patterns of resource distribution and economic inequality, it follows that the question of home ownership must remain as central to the analysis of social divisions and political conflicts. (23)

That is a very large claim indeed. It has been challenged by Harloe and there is nothing in that challenge with which I disagree. In particular Harloe points out (24) that when Saunders says: "One's class location does of course set limits upon one's consumption location...but it does not determine it", that 'setting limits' is precisely what determine means here. As an early and rather more straightforward Weberian piece stated:

> It is through the mechanism of housing that major life experiences conventionally associated with occupational class are determined: housing's relevance for stratification is not just an index of actual achieved life chances, but as the means by which the inequalities of the occupational structure are transferred into the wider social structure. (25)

Indeed the Paynes' efforts at developing the notion of 'housing status group' (26) seem rather more convincing as a Weberian method of comprehending this area than Saunder's very rigid demarcation of production from consumption. Harloe's critique of this approach closely resembles the general position advanced by Ball. (27) In general, but not always, this is a capital logic approach. It asserts that housing issues cannot be separated from the relations of production, although it has not usually employed the notion of 'reproduction of labour power' as a linking concept. Sometimes it does go further. Ball's comment that:

> It is clear that the economics cannot be divorced from the politics of housing provision... This is as true for private tenures like owner occupation as it is for state owned council housing. To understand the economic interlinkages involved in housing provision therefore, it is necessary to consider the political process in more detail

is a good prescription. He then starts us off on the right lines:

> Rarely does political pressure over any issue derive from an homogeneous block like the working class. Instead it comes from amorphous groupings, the constituent parts of which vary over time, and frequently those parts are not formally linked nor recognise their common in-

> terests. Such groupings, moreover, often coalesce around specific issues rather than broad long term factors. (28)

This argument is then pursued in a way with which insofar as it goes, I agree and would note particularly the very different definition of tenure employed from that by the same author which I have criticised earlier. (29) Why did I then say it was unsatisfactory as a way of setting the terms of an account? Well, on reflection that was a somewhat exaggerated statement, but still let's look at the implications of 'social being determines consciousness'.

What is interesting about the new quasi-Weberian account of this area is the most unWeberian methodological character of its, rather meagre, empirical base. When Saunders suggests that:

> ...it does therefore seem plausible to suggest that ownership of housing may be very significant in shaping people's political values and in structuring political alignments as well as in generating a distinct 'owner occupation' interest which no government can afford to ignore (30)

the demonstration, as opposed to assertion of plausbility, must lie with face-sheet analysis of voting behaviour surveys, and, at a slightly more developed level, of say 'The British Social Attitudes Survey'. (31) Personally I see nothing wrong with this, <u>as a starting point</u>, but it has very little to do with verstehen, with any sort of explanation which meets the criterion of adequacy at the level of meaning.

Marxist accounts, and production centred accounts of whatever form in general, are rather better here because of their relationship with a series of historical studies, but what about what is happening now? Earlier Weberians (Rex, Dennis, etc.) were rather good at going and looking at this. I seldom agree with the conclusions they draw, but the research effort was there. What we really do lack is any sustained and general (that is to say looking in a lot of different places; at different ways in which changes are working out), ethnography of these issues. It is a case of come back the community study - much, if by no means all - is forgiven.

There is a well known distinction between 'class in itself' and 'class for itself', an equally well known phrase about making history but not in circumstances of own choosing, and another about the best way to understand the world being to change it.

How do people perceive the social world and then act out their perceptions? Or, rather, what do they perceive, as the basis for subsequent action? Lives are not partial, in perception any more than in determination. People in Shipcote/Saltwell are experiencing changing relations in production and reproduction. Their perception will include both these process sets. They are also perceiving action, including their own action. The largest and most active housing issue group in Gateshead is that of working class owner occupiers in this area demanding the restoration of grants - something which brings them into direct conflict with the logic of fiscal crisis for a marginalised group. These people are actually starting to act collectively, as a collectivity, rather than as an aggregate of similar individual interests. (It is by the way worth noting the importance of spatial propinquity as the basis of these actions). In the sense of the word 'determine' suggested above, social being is determining consciousness which is determining action. How do we understand this?

In attempting to frame a problematic, a general definition of the nature of the issue and of approaches to it, I have been much assisted by reading two recent contributions by Fred Gray and Mike Ball. I find myself in virtually complete sympathy with Gray's review of these issues in a chapter entitled 'Owner occupation and social relations'. Let me quote a precised version of his conclusion.

> Owner occupation does not have a single independent effect on social relations. The tenure itself is not monolithic, but varies spatially and temporally. Neither is it independent of society wide economic, social and political processes. Similarly, the social relations of home owners are not homogenous, and are not the consequence of owner occupation per se. ...Longer term economic change, the roller coaster of the capitalist economic system, has also affected the role of owner occupation in determining social relations. During the periods of prosperity shared by the majority of the population many people have had a better material life and a freer range of social relationships through home ownership, particularly in contrast to tenants of state housing as it presently exists in Britain. The reverse is also true. Economic crisis increasingly infringes on the ability of

> the tenure to provide people with satisfactory housing or to produce a stable social structure. Under these conditions the potential grows for housing or wider political struggle by home owners. Yet in turn, this potential may only be realised through political awareness and activity arising though wider class consciousness. (32)

It seems to me that the historical example of Saltwell/Shipcote is illustrative here. The area was built in a period of local capitalist boom between 1880 and 1910. It was constructed in consequence of industrial growth and development to reproduce a necessary labour force. Its present problems can quite properly be located in relation to basal change at the level of the spatial organisation of the system of production. In this respect, although I agree with the general tenor of the remark, I would go further than Ball when he observes:

> Instead of simply delimiting a means of consumption, housing tenure is associated with historically specific relations of provision. A series of social agencies are associated with housing provision in any tenure form. Their existence is not a necessary consequence of the tenure itself, but the product of long historical struggles. The interrelation of these social agents determines the contemporary characteristics of a tenure form like house price inflation or jerry built high rise council blocks. To ignore these social agencies leads to a failure to analyse the causes of particular characteristics of housing tenures that are felt to need change. It can also lead to a misspecification of the central problems with a housing tenure. One characteristic misspecification is to place sole blame on the state and its subsidy policies for all the ills of housing provision. In theoretical terms the traditional tenure focus of housing policy is trapped in the spheres of distribution and exchange. (33)

Put crudely Shipcote/Saltwell as a social space and set of physical constructions relating to that social space was the product of a particular set of historical class production political relations. The 1960s-70s social and physical reconstruction of the locale which involved a major tenurial change, were the product of another set of class production political relations. The present situation is that of a crisis in general which has its origins in a production crisis but which is expressed in all spheres including that of the nature of national and local state housing policies and their consequences. Fred Gray suggested in 1982 that:

> Under these circumstances, low income owner occupation may indeed become dysfunctional for capitalism. The legitimacy of private property is, in theory, threatened for in some inner city areas it ceases to be seen by people living in obsolete dwellings as a socially useful good, but more as a disadvantaging drain on financial and other resources, particularly in comparison to relatively advantaged tenants in good council houses. (34)

I think that the second half of this is wrong although I would have, and did, say exactly the same thing in 1982.

In fact the tenants in good council housing, who have almost exactly the same production relations as the Saltwell/Shipcote population, are experiencing exactly similar pressures expressed through the reduction in local authority capital programme. Much of the good interwar council housing is in fact at the same stage in its designed for life cycle as Shipcote/Saltwell and is not receiving investment for the same basal reason, that is the socio-productive political peripheralisation of the population who inhabit it. However, I think that the first part of the quotation is correct. The housing market is not collapsing in Gateshead in the owner occupied suburbs but a part of the private system is in serious stress. The social implications are profound. What is going to happen?

Let me conclude very quickly. At the very least we observe it. In a time of changes we look at the changes. We may even feel obliged to join it. But we cannot look at bits. This is not a rejection of particularly in studies. Gateshead is clearly very different from some other places. It is a plea for some general empirical work which ought not to be abstracted empiricism but rather conducted within a problematic based on an action centred conception of the relationship between base and superstructure, production and reproduction, social being and consciousness, even if the relationship among these pairs is not very well sorted out.

NOTES AND REFERENCES

(1) E.P. Thompson, The Poverty of Theory, Merlin, London, 1978, p. 85.
(2) R. Williams, Problems in Materialism & Culture, Verso, London, 1980, p. 34.
(3) P. Saunders, 'Beyond Housing Classes', International Journal of Urban & Regional Research, vol. 8, no. 2, 1984, pp. 202-27.
(4) M. Harloe, 'Sector and Class: A Critical Comment', International Journal of Urban and Regional Research, vol. 8, no. 2, 1984, pp. 228-37. M. Ball, 'British Housing Policy and the House Building Technology', Capital and Class, no. 4, 1978, p. 85.
(5) A. Stewart, Housing Action in an Industrial Suburb, Academic Press, London 1981, p. 213 and subsequently.
(6) A. Friedman, Industry and Labour, Macmillan, London, 1977.
(7) D. Byrne, 'Just Hold on a Minute There', Capital and Class, no. 24, 1985, pp. 75-98.
(8) E. Mingione, Social Conflict and the City, Blackwell, Oxford, 1981.
(9) North of England County Councils' Association, State of the Region Report, 1981, p. 3.
(10) Gateshead Figures, Gateshead M.B.C. Quarterly Review of Unemployment, Sept. 1984 - Regional Figures, Dept. of Emp. Press Notice, 29.11.1984.
(11) C. Cockburn, The Local State, Pluto, London, 1974, p. 56.
(12) H. Lefbvre, The Survival of Capitalism, London, Allison and Busby, 1976.
(13) P. Anderson, 'The Antinomies of Antonio Gramsci', New Left Review, no. 100, 1977, pp. 5.80.
(14) C. Ungerson, and V. Karn, The Consumer Experience of Housing, Gower, London 1980.
(15) Ibid., p. xv.
(16) Ibid.
(17) Ibid., p. xvi.
(18) Ibid., pp. xvi-xvii.
(19) See N. Dennis, People and Planning and Public Participation and Planners' Blight, Faber, London, 1970 and 1972, for a developed account of a similar locale.
(20) M. Banim, work in progress 1985, Sunderland Polytechnic.
(21) P. Dunleavy, Mass Housing in Britain, Oxford University Press, Oxford, 1981.
(22) A. Stewart, op. cit., p. 216.

(23) P. Saunders, op. cit., p. 207.
(24) M. Harloe, op. cit.
(25) J. and G. Payne, 'Housing Pathways and Stratification', Journal of Social Policy, vol. 6, no. 2, 1977, pp. 129-50.
(26) Ibid., p. 132.
(27) M. Ball, 'Housing Provision and the Economic Crisis', Capital and Class, no. 17, 1982, pp. 60-78.
(28) Ibid., pp. 65-6.
(29) See Note 4
(30) Op. cit., p. 207.
(31) Social and Community Planning Research, British Social Attitudes: The 1984 Report, Gower, London, 1984.
(32) F. Gray in S. Merett with F. Gray, Owner Occupation in Britain, Routledge and Kegan Paul, London, 1982, Chapter 13.
(33) M. Ball, 'Coming to Terms with Owner Occupation', Capital and Class, no. 24, 1985, p. 24.
(34) Op. cit., p. 282.

Chapter Seven

HOUSING POLICIES FOR OLDER PEOPLE

Christine Oldman

Introduction

It might be argued that the grievous housing problems of the mid-nineteen eighties have less affected older people than other sections of the community. Monetarism applied to housing policy results in the provision of public money for housing only to those in special need. The elderly are seen, and have been seen for some time, as the deserving poor and thus it seems to be right that they should be the recipients of what meagre support is being made available within housing programmes. Certainly there is a great deal of talk, and even excitement about a range of housing initiatives in the 1980s for older people. The Department of Environment film 'Housing for the Elderly', first shown in 1983, concludes: 'Never has there been so much choice for old people'. The burden, however, of this chapter is that such assumptions about older people and their housing lives are, sadly, much exaggerated. Moreover it will be argued that the housing problems of old people which in reality are growing, not diminishing, have largely been converted or diverted into 'care' problems.

The debate about housing and old people is very lively. The supply of public sector sheltered housing, of the conventional type, certainly that managed by local authorities, has fallen off, but there is much talk about: extra care sheltered housing, privately built sheltered housing 'staying put' projects, community or dispersed alarm systems, about shrewd financial projects which will release the capital locked up in an old person's own house, etc., etc. Despite all this energy expended on older people very little has actually been achieved other than, perhaps, for those old people who can afford to buy a private developer's sheltered flat. All this buzz of excitement about the elderly may simply serve the purpose of filling a vacuum. Investment in housing not simply in the public sector but also in the private sector is so pared down that those involved in housing have to shout very loudly and cheerfully about the small help that is being offered to older people in order to offset the fact that nothing much is being done in other areas of housing policy.

Anthea Tinker's (1) latest report, clear and succinct as it is, quite well illustrates the two themes of this chapter: first that the reality of housing help to older people in the 1980s is a far cry from the rhetoric, and second, that the orientation of the help that is available is more to do with the provision of care services rather than a housing orientation aimed at a fundamental improvement in housing conditions. Tinker's report is a Department of Environment report, not a Department of Health and Social Security or even a joint publication, yet there is detailed discussion on 'initiatives' such as alarms or intensive home carer schemes or whatever, all of which are 'caring' policies and certainly traditionally have been personal social services departments' responsibilities. Admittedly there is a very strong argument for housing agencies to carry out welfare functions but there is no justification for this to be at the expense of investment and effort to improve the <u>housing</u> lives of old people. Quite simply housing policies for old people are not and have not been for a long time <u>true</u> housing policies - they are welfare or care policies. Why should one be so aggrieved at this? It is the task of this chapter to try and give some sort of answer to this question. The prelude to the answer is that old people are in some way blamed for being old. All the effort is on them themselves, in, for example, the provision of a commode because of an inability to climb stairs. There is no real attempt to radically improve basic housing conditions. It is only this second emphasis which is going to lift Britain out of its housing crisis.

Rather than old people being the one group that is protected from massive reductions in housing and other closely related public expenditure the reality is that the housing lives of old people serve as a good example of what is meant by the housing crisis of the 1980s. From the particular one can move to the general. In other words by providing an analysis of what is happening in housing terms to old people one can very well illustrate all the facets of the wider British housing crisis. Certainly the tendency to avoid the housing problems of old people and to go for a 'caring' situation finds a parallel in housing policy at large since the mid-nineteen seventies in the sense that there has been a shift towards greater selectivity. The 1972 Housing Finance Act prefaced a change in policy from one of subsiding homes to subsidising individuals by providing them with welfare payments.

> Traditionally subsidies related to the cost of providing (local authority) dwellings but now the bricks and mortar subsidy has been changed into a means tested form of income support for the poor. (2)

The aim of this chapter will be to defend the claims in the introduction that: (i) old people have not been particularly protected from misfortunes of British housing and; (ii) that policies have been more 'caring' policies than fundamental housing policies. The chapter will defend these claims by:

- Summarising the empirical data relating to the housing conditions of old people;
- Examining the contribution of sheltered housing to the housing lives of older people;
- Evaluating the changes affecting the older people in housing policies since 1979.

A Housing Profile of Older People

Older people's housing conditions are poorer than the majority of those who are not old. There are three major reasons for this:

- Older people tend to live in older houses;
- More old people than younger people live in privately rented accommodation which contains a disproportionate amount of deteriorated property;
- There are considerable numbers of low income home owners.

These three factors are related and the purpose of this part of the chapter is to discuss both the age of elderly housing and tenure and the relationship between them. Firstly, however, it is necessary to remember that though older people are satisfactorily housed, those who are not are proportionately greater in number than those poorly housed in the community at large. There is now a considerable amount of empirical data to support this last point, the most notable and recent sources being the 1978 National Dwelling and Housing Survey, (3) the 1981 Census, (4) the 1981 English House Condition Survey (5) and the 1982 General Household Survey. (6) The table below extrapolates from two of these sources to show facets of the housing disadvantages of older people.

Table: Tenure and Amenity of Older People's Homes

	Elderly Households		
	One Adult 60 or Over %	Elderly Couple* %	All Households %
Tenure:			
Owner Occupied	38	53	55
Local Authority+	42	36	32
Private Rented	15	9	3
Rented from HA	3	1	2
Amenities:			
Lacking Indoor Toilet	7	4	2
Without Central Heating	49	46	40

Source: 1982 General Household Survey
National Dwelling and Housing Survey 1978

* one of both persons aged 60 or over
+ includes renting from a New Town Corporation
HA Housing Association

The table presents the tenure breakdown of older persons' households. Tenure is, of course, of great significance in understanding both the housing conditions of old people and housing policies aimed at old people. Almost half of old people are owner occupiers. Moreover the numbers of elderly people in this tenure are increasing; firstly because people who reach retirement age now are more likely than before to be home owners and secondly growing numbers of older council tenants are buying their houses.

There are a number of interesting issues involved in elderly owner occupation which are indicative of wider housing policy themes. There has been the assumption, at least on the part of local authority housing managers, that those in owner occupation are satisfactorily housed. Certainly the Leeds University Evaluation of Sheltered Housing (7) showed that elderly owner occupiers were discriminated against in the allocation of sheltered housing. Interestingly the 'problem' of the elderly owner occupier has only recently been put on the housing policy agenda. Yet Cullingworth (8) as early as 1969 in his exhortations to policy providers to adopt a more generous definition of housing need discussed owner occupiers' problem - the maintenance of over large houses and gardens, etc. His words were largely unheeded, but there were more powerful ways of making sure the elderly owner occupier did get on the political agenda. Firstly there was the realisation that increasingly the country's social security budget was being used to finance, albeit in a meagre fashion, house maintenance. David Donnison, when Chairman of the late Supplementary Benefits Commission, said that the supplementary benefits system was "wholly inadequate to prevent the elderly's housing stock deteriorating." (9) Considerable numbers of pensioner owner occupiers are in receipt of supplementary benefit and each General Household Survey shows greater numbers living at the margins of supplementary benefit. The second reason for the discovery of elderly owner occupiers was the publicity given to the 1981 English House Conditions Survey. The survey, published at a time when expansion of owner occupation was being actively encouraged by the Government, highlighted the state of disrepair and neglect in that tenure which supposedly enshrines the values of self-sufficiency, independence and price in possessions. The problem was found to be most acute in pensioner householders. It was and is supposed that the problem is most serious amongst this section of society because old age brings with it incapacity and a consequent lack of energy or motivation to cope with house maintenance or repair.

The English House Condition Survey like previous surveys demonstrates clearly the links between age of housing and poor housing conditions. Elderly owner occupiers are more likely to be living in pre-1919 housing than younger owner occupiers. Measures of housing conditions have been inadequate in that they measure that which is not easily quantifiable. Consequently there was an optimism that housing problems had been greatly ameliorated because the two favourite indicators, basic amenities and unfitness for habitation, gave different, more favourable scores than the two previous English House Condition Surveys. Anthea Tinker writes:

> The recent housing surveys show that housing conditions have improved for the elderly as for the rest of the population. (10)

The numbers of elderly households in dwellings lacking one or more basic amenit
fit for habitation has indeed been reduced but there are other measures of housing need amongst which state of repair is the most significant. The 1981 English House Condition Survey highlights the state of disrepair in the nation's housing stock: it shows that 41 per cent of elderly owners are living in 'poor' or 'unsatisfactory' houses (unfit or needing more than £2,500 on repairs) compared with 22 per cent of owner occupier households with a younger head.

The absence of one or more basic amenities is not an adequate measure of housing condition but it is an important indicator of the housing disadvantage of older people. The number of houses without basic amenities is reducing but still, as the previous table shows, for one of the basic amenities at least, the elderly suffer more than the rest of the population. The problem is more acute for very old people. The 1981 census showed that 30,000 aged 85 or over live in homes that have no indoor toilet. This surely is housing deprivation.

Tenure is a key factor in discussing housing disadvantage in Britain. The table shows the number of old people living in public sector housing. Yet its contribution to improving the housing lives of old people is quite considerable. The housing problems discussed above most affect those old people living in the private sector either owner occupied housing or privately rented accommodation. Although the problem some council accommodation have been well documented the 1981 census shows that three in every ten lone old people live in purpose built flats or maisonettes. Some council housing, reflecting the vagaries of housing policies since 1919, is of superb quality and surpasses much privately built housing. From 1945 to 1960 small accommodation was 10 per cent of the total, from 1966 to 1971 it was 27 per cent. Since 1970 almost a third of all new building by local authorities has consisted of one bedroom accommodation, most of which according to Tinker (11) was probably for occupation by old people. Well managed, well maintained, well heated council accommodation represents a considerable benefit to old people.

The tenure which receives no financial support from the state other than welfare payments in the form of housing benefit to individual tenants is the private rented sector. In the second half of this discussion of housing policies directed at elderly people I shall show that the plight of old people in this sector has been largely unnoticed. The sector may well be declining inexorably but as the table shows 15 per cent of lone elderly households live in it.

It is not enough to detail housing facts and figures pertaining to those of retirement age. It is necessary to reflect on the consequences of living in houses that are cold and in a state of disrepair. The essence of the problem for so many old people is that their houses are simply not warm enough. Ill fitting windows, slates off roofs, crumbling pointing will all result in a loss of heat. Warm houses are important for those with chronic conditions, whose mobility is restricted in any way. However as I will attempt to demonstrate the response has been to focus on the special needs and characteristics of old people rather than the housing conditions which has helped to exacerbate the chronic conditions from which they suffer. It is self-evident that housing is of central importance to people who by virtue of having no paid employment to take them out of their homes and/or by suffering some condition which prevents them from getting out of their front doors. Yet there is inconvertible evidence that older people endure poorer housing conditions than those who are able to leave their own four walls. Social and economic processes 'retire' people at 60 or 65 and so consequently mean income and mean health confine older people to their homes. However there is a danger when making this point of perpetuating the ageism which pervades housing and social policies in this country. The elderly are not alone living out virtually their entire lives at home. For example the young handicapped male, or the young woman with two preschool children can feel miserable with their living circumstances. The common thread is that the deficiencies of one's home becomes rather more compelling if one spends most of the twenty four hours of the day inside it. The elderly, and other low income groups who spend much of their time in the home. The points to be established here are that the elderly are disproportionately affected by the current housing crisis, and that their main requirement is for decent, warm, suitable, ordinary accommodation at a price within their means. The main thrust of policy, however, as subsequent sections demonstrate, has been to provide high quality specialist accommodation for a very small proportion of the elderly population, with very little being done for the great majority of them.

Sheltered Housing

The concern of this section is to comment on housing policies directed at older people up to the election of the Thatcher government in 1979. Essentially this commentary will examine the development of sheltered housing since this form of provision dominated thinking about housing for old people more or less up to 1979. The argument to be presented here is that sheltered housing was developed very much to the exclusion of any other housing initiatives. The sheltered story continues after 1979 but it is a convenient point at which to break since the election of the Conservative administration in that year represents some sort of watershed in British housing policy.

Sheltered housing has been hailed as the greatest local authority success. It is a little difficult to establish quite what is meant by the claim. It may be cynical to suggest that its success is by default. The development of sheltered housing has to be compared with the lack of activity in other spheres of public sector housing. The process known as residualisation of council housing (12) was under way by the nineteen seventies.

Enthusiasm for debates about sheltered housing, hectic discussion both for and against sheltered housing continues unabated. Conferences on the subject still draw crowds. The energy expended on sheltered housing far exceeds its numerical significance. It must be stressed over and over again that sheltered housing only provides for a very small proportion (5%) of the elderly population. (13)

In the previous section of this chapter the tendency for local authorities to respond to the demographic trend towards small and single households was referred to. Anthea Tinker (14) notes that since 1970 a third of local authority new build was one bedroom accommodation. She also speculates that this would be allocated to old people. There are two points to make here. One is that local authority new building during most of that decade other than a brief mini boom in the mid-seventies was decreasing. The second is that the elderly are bidding with many others for small accommodation.

There is often considerable difficulty in deciding what precisely sheltered housing is. Butler et al approached the problem thus:

> We define three elements as distinguishing sheltered housing for the elderly from other categories of housing, namely: a resident warden, an alarm system fitted to each dwelling, and the occupancy of dwellings being restricted to elderly persons. Another typical, but not universal, feature is that dwellings are grouped on one site whether 'on the ground' or in blocks of flats. Sheltered housing has usually been purpose built but some has been created through the conversion or adaptation of existing housing. (15)

Housing policy in the interwar years and in the immediate postwar era was not concerned with the needs of older people. The overriding objective of the post 1945 period was to alleviate the chronic housing shortage of the time. The effort in the public sector and then later, in the 1950s in the private sector, was to provide general needs family housing. However during the fifties and the sixties governments actively encouraged local authorities to consider specialist accommodation for old people. It is interesting that as early as 1954 the potential for sheltered housing in providing allegedly cheap 'caring' accommodation within the community was recognised. The Philips Committee in 1954 recommended that elderly people:

> should as far as possible continue to live as members of the community. With this end in view we consider that it is important that special housing of various types, adapted to the needs of old people but not isolated from the rest of the community should be provided. (16)

It is sometimes forgotten by commentators that arguments about the role of sheltered housing, which has often been documented in the housing and personal social services' press, are not new. It is understandable that, as resources for social spending from the late seventies and increasingly now in the nineteen eighties dwindle, discussions about the silting up of sheltered housing (17) and the moves to push it further along the care continuum by the provision of extra care schemes such arguments would appear recent. But they are not. Essentially the provision of accommodation designed specially to meet the alleged special needs of older people is a response to the problems of old age, not a response to the housing crisis that so many old people find themselves in. An analysis of the many Ministry of Housing and Local Government design bulletins shows this emphasis on a caring rather than a housing orientation quite clearly. For example 'More Flatlets for Old People' (18) published in 1960 advocated the building of flatlets because they were easy to manage and made possible the provision of a warden who was available to help in case of emergency. A year later the Ministry of Housing and Local Government and Ministry of Health jointly published a circular which stressed cooperation between the various services involved with the elderly. (19) It suggested ways in which the existing services could be improved so that adequate provision for all the various needs of old people could be met. It repeated the belief that old people prefer to live an independent life for as long as possible. The importance of the warden was stressed and her role outlined.

> Details of arrangement differ but often a warden undertakes to clean the common room, bathroom and WCs, landing and stairs, and attend to the central heating; answer the emergency bell and apply for services needed by the tenants, such as home helps, meal services and supplements to pensions. In addition, many wardens help with household tasks such as hair washing and bathing. They also draw pensions, shop and cook in bad weather or illness and organise social or special parties in the common room. (20)

It was not until the end of that decade that the celebrated blueprint which has dominated all thinking about housing the elderly appeared, thus blinkering policy makers from other possible approaches. This was Circular 82/69 which remained on the table until 1981 when the Thatcher administration withdrew it, along with the housing cost yardstick. The theme of early circulars was that:

> The purpose which underlies the design of housing for the elderly is the provision of accom-

modation which will enable them to maintain an independent way of life as long as possible. (21)

The circular was a detailed design guide. It did consolidate earlier advice on housing the elderly but it introduced the notion of two sorts of schemes and, therefore, by implication two sorts of elderly - category 1 and category 2. Category 1 schemes, generally bungalows, were to be 'self contained dwellings to accommodate one or more old people of the active kind'. The tenants, it was envisaged, should be 'couples who are able to maintain a greater degree of independence, who can manage rather more housework and who may want a small garden'. The more costly category 2 schemes would usually consist of flats and have a wider range of additional services; a common room, laundry, public telephones, etc. These should be for 'less active old people, often living alone, who need small and labour saving accommodation'. Optional extras within such schemes might include a warden's office and a guest room.

Sheltered housing developed quite dramatically during the 1960s and 1970s. The 1974 Housing Act had the effect of boosting housing associations' contribution in this field with their remit to respond to 'special needs'. By the start of 1970 the number of units of sheltered housing in England and Wales had risen to almost 100,000 compared to 21,000 in 1960. By the middle of 1983 there were 323,600 units of sheltered accommodation in England and Wales. (22) It has been noted, however, that enthusiasm for sheltered housing varies quite considerably. Both the Leeds University study (23) and the Oxford Polytechnic survey, (24) carried out on behalf of the Department of Environment, show a wide range of provision per thousand of the elderly population living in any one local authority's jurisdiction. Despite the fact that sheltered housing has a caring ambience it is a housing provision developed and managed by housing agencies. It is not subject therefore to the statutory norms which operate in the personal social services field. The obvious implication, therefore, of the semiautonomy allowed to housing agencies is that old people may be lucky or unlucky if they require a sheltered unit depending on where they happen to live.

It has been noted that a considerable amount of energy is devoted to housing the elderly on the part of housing agencies particularly as financial support is being withdrawn from other areas of housing endeavour. Demographic changes over the second half of the twentieth century have also prompted local authorities to respond to the needs of large sections of their communities. The national average percentage of people above retirement age is 18 per cent. The total number of elderly will not rise appreciably in the future. What is most significant now is the decline in the number of young elderly and the rise in the number of very old. For instance those over 85 will increase from 5 per cent in 1977 to 8 per cent in 1991 and remain at that level in 2001. (25) Local authorities not only are faced with large elderly populations but also with the phenomenon of underoccupation. Social Trends, 1980, quoting the General Household Survey, reported that 67 per cent of individuals and 80 per cent of couples aged 60 or over had accommodation with one or more bedrooms in excess of their 'requirements'. The highest levels of underoccupation are to be found not surprisingly, among the ranks of elderly owner occupiers. The Leeds University research found that elderly council tenants were often encouraged to move into sheltered accommodation to release a house for a family on a local authority's waiting list. Owner occupiers, of course, are free from state persuasion of this sort. The issue of underoccupation is a controversial one. There is evidence that the provision of sheltered housing in the public sector is a response to underoccupation. Yet as the owner occupied sector continues to grow, housing requirements or housing need will give way to financial motivations. As many housing writers have commented, the present system of demand subsidies to home owners actively encourages underoccupation. Housing choice thus comes to reflect financial advantage rather than requirements for living space.

The caring element in sheltered housing has, it has been shown, a long history. Sheltered housing is a physical embodiment of community as opposed to residential living. It is rare that academics have much influence over the evolution of social policy. Yet Peter Townsend's book 'The Last Refuge' published as long ago as 1962 has probably been responsible for the development of sheltered housing. In this book he painted a gloomy picture of those who spent their last years in institutional care.

> They (the residents) are subtly orientated towards a system in which they submit to orderly routine, lack creative occupation and cannot exercise much self-determination. (26)

He advocated an extensive sheltered house building programme which he hoped would largely replace institutional care for the elderly. He called for a provision at the rate of 50 dwellings per 1,000 of the population aged over 65. Since Townsend published his book, however, much has changed in the social spending sphere. Sheltered housing has become an alternative to residential care but much hangs on the word alternative.

> Explicit in Townsend's use of the word was that sheltered housing should be <u>different</u> from residential care; the development of sheltered housing, he argued, would reduce the necessity, for Part III accommodation. Others use the word 'alternative' in another sense, referring to sheltered housing as being complementary to or similar to residential accommodation. (27)

Sheltered housing now is complementary to residential care. In the latter half of the nineteen seventies much interest was being shown in what is variously called extra care sheltered housing, very sheltered housing or category 2½. This interest has come about very largely as a result of reductions in social service budgets. The popular legend was that tenants in sheltered housing were ageing and thus becoming more dependent and, at times, putting intolerable burdens on the warden. No help or insufficient help was available from social service budgets. The Leeds University research attempted to rigourously test this assumption. We found much evidence of ageing and dependency. However this overall conclusion has to be put into context. Firstly the levels of dependency in sheltered housing schemes were not greater than that found in the community and the support given to sheltered housing tenants was much greater. Secondly the very existence of a warden promotes dependency as well as satisfying it. The Leeds conclusions were, it is fair to say, not always accepted by those involved in housing.

Certainly extra care schemes have developed as a result of the belief that conventional sheltered housing was becoming unmanageable. There are now, as will be shown in the next section, some extra care schemes in operation. It seems possible that management problems will not be eased.

> It is well established in social policy that a new service or provision not only satisfies and meets a need but also reveals or generates

> further expectations, needs and demands. The shift in emphasis to extra care risks creating just another institution to be added to the continuum of care, creating further discussion, disagreement and ambiguity at the boundaries of each type of provision. (28)

What is interesting about the emergence of extra care sheltered housing is that it continues and strengthens the caring orientation towards the housing needs of old people and further deflects attention away from the appalling housing conditions of many old people which result from their economic disadvantage in society. Until very recently with the emergence of private sector sheltered housing it has not really been possible to talk about a true demand for sheltered housing. Instead with developments such as extra care an old person's 'need' for sheltered housing is assessed not by they themselves but by others - doctors, welfare professionals and housing visitors.

This section is concluded by a critique of sheltered housing. The Leeds University research has been taken to task by some (29) for being critical of sheltered housing as a housing policy for old people. Such people say that enormous waiting lists for sheltered housing do not suggest antipathy towards sheltered housing on the part of the elderly. We were not critical of sheltered housing. We had hoped to be interpreted as saying that we felt the adoption of sheltered housing as the 'ideal solution' to the housing problem of the elderly. In 1976 the government itself urged local authorities to move away from the Circular 82/69 straitjacket. It urged authorities to provide:

> a full range of small bungalows, flats and flatlets designed for old people; some in which they can be fully independent (though with neighbours at hand in case help is wanted); others in which some friendly help is available in the person of the warden; others still in which provision can be made for some communal services in addition. (30)

It could be argued, and most certainly would be argued by this present government, that the sheltered housing mould has been broken and that there is a great deal more flexibility of provision. In the next section I shall argue that <u>real</u> housing choices available to old people are very limited and will continue to be so as long as housing investment remains at its pitifully low levels. Private enterprise can only rescue a limited number of old people from their housing plight.

Many claims have been made for sheltered housing which empirical data suggest may be unfounded. There does seem little doubt that sheltered housing is a housing success in the sense that tenants' housing conditions have been greatly improved. The clear theme to emerge from the Leeds research was that improved housing was the most salient feature of sheltered housing for its tenants. Alarm and warden services were far less so.

> What is not altogether clear is why somebody living in poor housing conditions should be seen as a candidate for a form of specialised housing, when apparently their requirements could have been met in other ways - either by home improvements or a move to better quality housing. (31)

The proponents of sheltered housing have claimed that a move to sheltered housing prolongs life, reduces loneliness and fosters independence. However one in five of the Leeds sample would have preferred not to have moved from their previous home, many reported themselves to be isolated, certainly many studies of sheltered housing have shown the communal facilities to be underused. The move to push sheltered housing further along the care continuum would appear to threaten the notion of independence and the rights enshrined in tenancy as opposed to the lack of rights embodied in the notion of resident or client.

The Leeds University research was not sheltered housing bashing. It was quite evident that very many old people were happy living in sheltered housing. The danger, however, is of making generalisations on behalf of all old people. This point is made admirably by Muir Gray:

> Elderly people do not differ because of their age which ranges from 65 to 110. Their difference is forced upon them by virtue of low income and immobility. (32)

The most important claim and that which most fits the theme of this chapter and indeed of this book is that sheltered housing is a satisfactory response to the special needs of the elderly. Emphasising the special needs of old people does them a disservice. Many people find it very hard to accept that the elderly perhaps do not have so very many special needs. It will be said - 'Of course they do, old age brings infirmity'. The facts are unassailable. The 1982 General Household Survey shows that 5 out of every 10 elderly persons experience a limiting illness. The truth of such a statement is not denied. The question is why should these special needs mark old people off from the rest of the population and lead to the development of segregated housing? There are just as many disabled younger people as there are disabled people over retirement age. There is no reason at all, other than of course political ones, why housing should not be designed to suit a wide range of physical needs - why, for example, could we not have more single storey accommodation? Why could we not have warmer, better insulated homes? The physical characteristics of old age are used to disguise the social and economic disadvantage of old age. Old people do have needs; of course they do, so do lots of other people. These needs could well be met by extensive home improvement policies instead of piecemeal help from welfare services. Housing solutions are rarely thought of. The elderly's special needs are exacerbated by the ageism evident in our society. Anthea Tinker admirably summarises the complex debate about special needs. She questions the need for special housing of any kind.

> It may be that the only way to get provision for disadvantaged groups is to make it easily identifiable (for instance, sheltered housing can be seen and counted whereas ordinary small houses interspersed amongst family houses are less visible), or to get better provision (for instance higher standards for the disabled). But we may do a disservice to a group by making them special...It is significant that the trend in other areas of social policy, like educational provision for the mentally and physically handicapped, is away from special provision and towards incorporation of special groups in normal forms of provision. (33)

Developments since 1979

It is not always wise to select a date as a peg upon which to hang a theme. The disillusionment with sheltered housing, experiments to shelter old people in their own homes and cutbacks in housing investment all preceded the election of the Thatcher administration in 1979. However that year is a convenient one to continue a discussion of whether the elderly have been singled out and protected as a group that requires housing support.

The 1980 Housing Act was of course a significant piece of legislation. It is rare, however, to find a specific discussion of how it affects the elderly. Of its three main constituent elements - the right to buy clauses, the tenants' charter and the new form of deficit subsidy for council house funding, it is the first which has received most attention. All these parts of the Act, however, do affect the housing needs of old people. The overall effect of the Act was to accelerate the process whereby council housing, indeed public sector housing as a whole (since public subsidised housing associations were affected by the legislation) should cater only for the poor and needy in society. But as the table earlier in this chapter shows large numbers of elderly people are housed in the public sector and so any adverse changes in the operation of public housing will greatly affect this group.

It may have seemed at first that the elderly would be protected by the Government from its plans for massive reductions in housing expenditure. The rhetoric, at least, was favourable. The year before the Act was passed Michael Heseltine, the then Secretary of State for the Environment, affirmed in a House of Commons debate the government's commitment to the elderly:

> We certainly intend to ensure that local authorities are able to build homes for those in greatest need and I have in mind especially the elderly in need of sheltered accommodation and the handicapped. (34)

We must evaluate that parliamentary statement. Firstly, is it factually correct and secondly, do the elderly benefit from being residualised, even paternalised? All the evidence (35) suggests that the pledge to preserve house building for groups such as the elderly has not been fully honoured. The new form of subsidies for council housing resulted in very low levels of new building, of all types including specialist accommodation for the elderly, and very much increased rent levels for tenants. Derek Fox (36) reports that since 1981 housing associations and local authorities between them have produced an average of 10,000 sheltered units. He also reports that very many authorities are looking at alternatives to sheltered accommodation, a quarter of all local housing authorities have diverted a high share of capital resources to central control alarm systems.

There are some problems in assessing the output of sheltered units since 1981. It could be said that an output of 10,000 units is impressive if compared with the sparseness of other new build activity. The philosophical commitment to sheltered housing on the part of local authorities and housing associations (particularly the latter - many associations have moved into the elderly field in order to keep going as viable organisations), is still there. There is still the belief that the housing needs of old people are best met by providing them with special and separate accommodation. This commitment, however, cannot now be realised as easily as it was in the nineteen seventies as a consequence of drastically reduced H.I.P. allocations.

Stressing, as the Government does that council housing should become welfare housing for groups marginal to society, such as the elderly, deepens the tendency to patronise and to stigmatise them. Peter Townsend has written persuasively on the theme of the social construction of old age. The effect of retirement, he argues, is not only to reduce income but also to cultivate dependency.

> While the institutionalisation of retirement as a major social phenomenon in the very recent history of society has played a big part in fostering the material and physiological dependence of older people, the institutionalisation of pensions and services has also played a major part. (37)

Malpass and Murie (38) and Malpass separately (39) have discussed the trend in social policy to move away from subsidising the physical structure that is the bricks and mortar of subsidising the individual in the form of welfare assistance. The high rent levels in council housing and fair rent housing have meant that those elderly who are unable or unwilling to avail themselves of government discounts with which to buy their own houses are dependent on the notorious Housing Benefit Scheme. Although at the outset of this scheme in 1982/83 some pensions did benefit, slightly, to the disadvantage of other nonelderly recipients, cuts in the scheme, with more in the pipeline, greatly affect old people. There are numerous accounts in the voluntary housing movement press of distress suffered by old people who have got themselves into serious rent arrears due to the administration difficulties faced by local authorities. Giro cheques both arrive late and can be underestimated. Traditionally old people do not get into debt. Housing association tenants are suffering the most as the hard pressed administrators of new scheme, the local authorities, have tended, perhaps understandably, to put this group at the bottom of their in-trays. The complex housing benefit supplement defeats both local authority officials and the elderly themselves.

We must now turn to the most controversial part of the Housing Act, 1980; the right to buy clause. The elderly were the focus of much heated discussion in the upper chamber of the Houses of Parliament and the Government was forced to concede that housing for old people should be exempted from the right to buy legislation. Although the Government accepts that some old people will require public sector housing at the same time it is active in encouraging and extending owner occupation amongst this group. It is noteworthy that the most repeated phrase in the Department of Environment's public relations exercise, the film Housing the Elderly is 'the private sector'. A similar battle to that fought preceding the 1980 Act was waged during the passage of the Housing and Building Control Bill through parliament. For instance the Government lost its clause giving tenants of charitable housing associations the right to buy their own houses but modified the legislation so that a tenant could be financially helped to buy a property on the open market.

It is a theme of this chapter that the elderly are as diverse in their needs, preferences and aspirations as the rest of us. Sadly, of course, housing and social policy, perhaps necessarily, attempts to iron out this diversity and to generalise about the elderly. The point is that some old people do want to avail themselves of generous discounts to buy their council houses (the 1984 Housing and Building Control Act extends the level of discount to 60 per cent of the purchase price for those with a tenancy of 30 years of more). For some old people, however, there may well be difficulties in either buying a council house or buying a property on the open market. Analysis of council house sales have shown purchasers to be older and poorer than other first time buyers. Middle aged couples therefore will be servicing a mortgage long into their retirement. Low cost home ownership, initiatives which reduce or defer costs in the early stages of a mortgage (unlike the traditional front end loaded mortgage) may well generate difficulties for old people as their income reduces.

The third plank of the 1980 Housing Act is the tenants' charter which was intended to answer criticism about the 'serfdom' of council housing tenants. It can only be a palliative measure during a period when high rents, council house sales and very low levels of housing subsidy all combine to make the lot of an existing council tenant a miserable one. It cannot be of much help to an elderly person, to receive a written statement of their council's allocation policies when their chances of transfer to a new flat are very remote.

Private Sector Sheltered Housing

Until the very end of the 1970s sheltered housing was not available on the private market. In a very short period of time there has been a remarkable emergence of a private sector in sheltered housing. (40) McCarthy and Stone, the pioneers, are no longer alone. A considerable number of private developers, housing associations either in partnership with builders or independently or part subsidised by Housing Corporation funding in the form of leasehold schemes for the elderly are entering the market. Such activity is now no longer confined to wealthy elderly people on the south coast of England. Some questions must be asked. Should we not welcome this development as representing a real widening of the choices available to old people? We have also, I think, to speculate (which is all that one can do) on why there was virtually no private sector sheltered housing before 1979/80. There are a range of possible answers to that question - there was no demand for sheltered housing on the part of old people; there was competition from the public sector - local authority and housing associations (this has very much been eliminated due to draconian cuts in public expenditure), private developers did not want the bother of dealing with old people; there was plenty of activity for them elsewhere, etc., etc. All these factors now appear to have disappeared. The Housing Research Foundation sponsored research at the University of

Surrey indicated a large and as yet unsatisfied demand for more sheltered units for sale. The study indicates a potential demand of up to 14,000 units a year. Wimpey has over 1,000 units in the pipeline on 30 sites, Barratt has at least 24 developments in progress while McCarthy & Stone, from a mere 38 started in 1978 expects to build over 700 units in 1985. Fielding notes 'the market in private sheltered housing could be worth up to £480 million a year by the end of the decade'. (41)

It really is not possible to evaluate private sheltered housing comprehensively. It is still in its infancy. However it my be useful to make some initial observations. Real choice for the elderly is to be welcomed. Private sheltered housing may serve the needs of those elderly owner occupiers who can realise sufficient capital for the purchase of their own houses. It is a provision which of course excludes many low income elderly home owners whose houses are not adequately attractive capital assets. Most of all it also excludes the other half of Britain's elderly, the nonhome owners. Sheltered housing as an ideal solution to the housing problems of the elderly has taken a severe knock in the public sector as a result of cutbacks in housing spending. It will not live on in the private sector or in partially subsidised form as represented by the Housing Corporation Leasehold Schemes for the elderly. Sheltered housing is a solution but devotion to it prevents experimentation with other options. Rose Wheeler argues not necessarily for less sheltered housing but for a housing strategy that will incorporate a range of provision:

> Complete replacement of some housing stock, and new building to include a choice of tenure and house type, including small, single storey homes, will be important features of any strategy...Mixed estates, and the use of infill sites for single storey accommodation, will help to stem the creation of elderly ghettoes, or age segregated accommodation. Housing for elderly people which is grouped together may contribute to an unnecessarily negative image of old age. (42)

Already some anxiety has been expressed about private sheltered housing. Firstly it seems likely that housing associations catering for the less wealthy elderly will be forced to compete with private developers and hence force up the price of sites. The literature of sheltered housing is dominated by discussion of management problems which arise when tenants become increasingly dependent and other solutions such as residential care are unavailable. Age Concern, for example, has expressed concern that problems will occur when an elderly leaseholder becomes ill and infirm. Many private leasehold schemes incorporate clauses which mean that when elderly people are assessed as not being able to look after themselves they will be forced to sell up.

Staying Put

If private sheltered housing is flourishing public sheltered housing is not. Doubts about its costs and tenants' privileged position were being expressed in the late nineteen seventies. These doubts have deepened since. Attention has now turned to what I shall call 'staying put' policies.

Staying put is a catch all phrase that can mean almost anything. With a capital S and a capital P it is a particular initiative, namely the Anchor Housing Trust experiment to use combinations of local authority improvement grant, building society finance and Anchor Housing Trust guidance and counselling in order to help old people improve the physical fabric of their homes. (43) Alternatively staying put instead of having a home improvement orientation can have a caring orientation and can refer to all those attempts to provide all or some of the alleged benefits of sheltered housing to old people in their own homes. Anthea Tinker's report 'Staying at Home' is representative of this particular orientation. It is ironic perhaps that the whistle has been blown on the policy of decanting old people from their own homes much later than the move away from clearance and new building to rehabilitation for the rest of the population.

Like sheltered housing, staying put in all its manifestations is to be welcomed. Certainly many old people do not want to move house. However the danger is, as with sheltered housing, the commitment could become single minded and exclude other housing possibilities. The elderly require a range of housing options not just one 'ideal solution'.

The 1981 English House Condition Survey identified great problems of disrepair in the houses of elderly owner occupiers. There has already been limited response to this problem by the Government and there is likely to be more. The long awaited Green Paper (43) on Home Improvement Policy, finally published in May 1985, stresses the need for targetting improvement resources to those in special need, including the elderly. Wheeler comments:

> Home improvement policy through the 1970s and early 1980s has failed to recognise the particular difficulties low income owners have in maintaining or improving their homes...The 'first come first served' basis of grant distribution in most local authorities has above all, penalised older people. (44)

The inability of old people to maintain and repair their homes is blamed on their decrepitude, their inability to cope with bureaucracy and complexity of grant procedures. As Wheeler so cogently argues it is not the elderly people who should be blamed but the grant system itself. It has defeated many a younger person. The greatest impediment to home improvement identified by the research evaluating the Anchor Housing Trust's exercise, Staying Put, was lack of finance. (45)

The elderly owner occupier has come in for a great deal of attention recently because of the recognition of the potential capital 'locked up' in the bricks and mortar of an old person's home. Building Societies are increasingly willing to give interest only mortgages to elderly home owners for improving, repairing or adapting their homes. Remortgaging, moreover, has another aspect. There is growing interest in schemes known as mortgage annuity schemes or home income plans enabling people over the age of 75 to take out a substantial mortgage with which to purchase an annuity from an insurance company to provide an income, and income which is not going to necessarily be spent on housing. Wheeler comments:

> A conflict may exist between policies which seek to develop the potential of pensioners' own homes to provide income, and policies which aim to encourage home equity to be reinvested in housing and its upkeep...What criteria should be used to decide the balance between private (home equity) and public funding for components and adaptations to elderly owner's homes? How far should elderly owner occupiers be expected to release home equity for other purposes, such as care, to reduce the call on public resources? (46)

Conclusion

This chapter has attempted to review recent housing policy developments for old people. The housing crisis is very real for many old people. We have shown in the previous section that elderly owner occupiers are now the favoured 'client' group in housing terms. Recent initiatives will however only affect the better off elderly owner occupiers. Recent housing policy for older people has very largely ignored the needs of those elderly living in the private rented sector, ethnic elderly, homeless elderly. Recent surveys have highlighted the growing dimensions of this last problem. (47) The public sector elderly tenant seems unlikely to flourish.

What is needed for old people, as for the rest of society, is intensive investment on housing. What is needed for the elderly as for everybody else is a radical overhaul of the present system of housing finance. What is needed for the elderly, as everybody else, is a system of subsidies which removes tenure financial advantages and disadvantages. The preferred tenure would then be the one that most suited housing requirements at any one point in the life cycle. The persistent trend in housing policy to divert attention away from the elderly's housing plight caused by social and economic disadvantage towards caring solutions such as sheltered housing or home improvement policies which attempt to modify the muddled apathetic thinking of old people only to serve to separate and stigmatise old people. In the absence of a massive injection of public funds into housing - and such funds seem unlikely in the immediate future - what is needed on the part of housing, health and welfare professionals is some joint action which attempts to give the elderly person some sort of genuine say in their housing careers.

NOTES AND REFERENCES

(1) A. Tinker, Staying at Home: Helping Elderly People, Department of the Environment, 1984.

(2) P. Malpass and A. Murie, Housing Policy and Practice, Macmillan, 1982.

(3) Department of the Environment, National Dwelling and Housing Survey, HMSO, London, 1979.

(4) Office of Population, Censuses and Surveys, Census 1981: Persons of Pensionable Age, HMSO, 1983.

(5) Department of the Environment, English House Condition Survey 1981: Part 2, HMSO, London.

(6) General Household Survey, 1982, OPCS, HMSO,

(7) A. Butler, C. Oldman and J. Greve, Sheltered Housing for the Elderly, Allen and Unwin, London, 1983.
(8) Ministry of Housing and Local Government, Council Housing: Purposes, Procedures and Priorities, HMSO, London, 1969. (The Cullingworth Report).
(9) A. Donnison, speaking at the Shelter conference, July 1978, cited in Age Concern Information Circular, September 1978.
(10) A. Tinker, The Elderly in Modern Society, Longman, London 1984, 2nd edn., p. 83.
(11) Ibid., p. 86.
(12) A. Murie, Housing: A Thoroughly Residual Policy, in D. Bull and P. Wilding (eds.), Thatcherism and the Poor, Poverty pamphlet 59, Child Poverty Action Group, 1983.
(13) D. Fox, Sheltered Housing Surveys 1984, Housing, vol. 21, no. 2, 1985, 17-21.
(14) A. Tinker, The Elderly in Modern Society, op. cit., p. 86.
(15) A. Butler et al, op. cit., p. 1.
(16) Ministry of Health, Report of the Committee on the Economic and Financial Problems of the Provision for Old Age, (Phillips Committee), Cmnd 9333, HMSO, London, 1954.
(17) Bytheway, B. and James, L.'The Allocation of Sheltered Housing: A Study of Theory, Practice and Liaison', University College Swansea, 1978.
(18) Ministry of Housing and Local Government, More Flatlets for Old People, HMSO, London, 1960
(19) Ministry of Housing and Local Government and Ministry of Health, 'Services for Old People', 1961. Joint circular 10/61 and 12/61.
(20) Ministry of Housing and Local Government, 'Housing Standards and Costs; Accommodation Specially Designed for Old People, 1969, Circular 82/69.
(21) Ibid.
(22) House of Commons, Hansard, vol. 48, no. 49, 15 November 1983.
(23) D. Butler, et al. op. cit.
(24) Department of the Environment and Welsh Office. Report on a Survey of Housing for Old People Provided by Local Authorities and Housing Associations in England and Wales, conducted on behalf of the DOE by Oxford Polytechnic, 198 .
(25) Tinker, The Elderly in Modern Society, op. cit., p. 12.
(26) P. Townsend, The Last Refuge, Routledge and Kegan Paul, London.
(27) A. Butler et al, op. cit., p. 14.
(28) A. Butler et al, op. cit., p. 141/2.

(29) For example, S. Donnison and D. Page, 'For the Rest of Their Days?: A Study of the Council's Sheltered Housing Schemes in Hull', School of Applied Social Studies, Humberside College of Higher Education, 1984.
(30) Department of the Environment and Department of Health and Social Security, Housing for Old People, HMSO, London, 1976.
(31) A. Butler, et al. op. cit., p. 39.
(32) J.A.M. Gray, 'Housing for Elderly People: Heaven, Haven and Ghetto', Housing Monthly, vol. 12, no. 6, 1976, pp. 12-13.
(33) A. Tinker, 'Can a Case be Made for Special Housing?' Municipal Review, no. 566, February 1977, pp. 314-5.
(34) M. Heseltine, cited in Municipal Review, no. 602 (April), 1980, p. 6.
(35) For example, Housing and Construction Statistics, Part 2, December 1983, Table 12.
(36) D. Fox op. cit.
(37) P. Townsend, 'The Structural Dependency of the Elderly: A Creation of Social Policy in the Twentieth Century. Ageing and Society, vol. 1, part 1, (March), 1981, p. 5-28.
(38) P. Malpass and A. Murie op. cit.
(39) P. Malpass, 'Housing Benefit in Perspective' in C. Jones and J. Stevenson (eds.) Year Book of Social Policy in Britain 1983, Routledge and Kegan Paul, London, 1984.
(40) N. Fielding, 'Growing Old in a Home That's Not a Home', Roof, vol. 10, no. 1, 1985, pp. 12-14.
(41) Ibid.
(42) R. Wheeler, 'Housing and Elderly People' in A. Walker and C. Phillipson, (eds.) Ageing and Social Policy: A Critical Assessment, Heinemann, 1985.
(43) Housing Improvement: A New Approach, HMSO, May 1985, Cmnd 9513.
(44) R. Wheeler, 'Staying Put: A New Development in Policy?' Ageing and Society, vol. 2, part 3, pp. 299-329.
(45) R. Wheeler 1985 op. cit.
(46) Ibid.
(47) A. Tinker, The Elderly in Modern Society, op. cit., p. 211.

Chapter Eight

THE WORSENING CRISIS OF SINGLE HOMELESSNESS

Chris Holmes

The housing needs of single people have never been properly recognised in Britain. Recent trends, however, are making the difficulties experienced still more acute. Demographic changes mean that there are increasing numbers of single people of working age wanting to form independent households. Yet savage public spending cuts have drastically reduced the number of new homes built for rent by local councils and housing associations. The private rented sector has continued its inexorable decline. Relatively few single people can afford to buy their own home - and rising unemployment has restricted this option still more, especially for young people. The inevitable consequence is that large numbers of single people now endure intolerable housing stress - enforced sharing and overcrowding, inadequate temporary shelter or literally no home of any kind.

Estimates drawn from the 1981 census data suggest that there were 8.62 million single people of working age in England and Wales alone. (1) Single people are no longer a small minority, but represent 3 in 10 of the working age population. Just over half of these single people are living with parents (or grandparents) in private households. Figure 1 shows the estimated living arrangements of the rest, including 15% of the total in private households other than those headed by parents or grandparents, a very substantial 17% whose living arrangements are not known (this will include concealed single person households), 14% living as independent single person households (including 4% sharing their accommodation) and 3% in communal establishments, such as hostels, common lodging houses and bed and breakfast hotels.

The most striking fact is that only 1,341,382 single people formed independent households - fewer

Figure 3: The living arrangements of single people

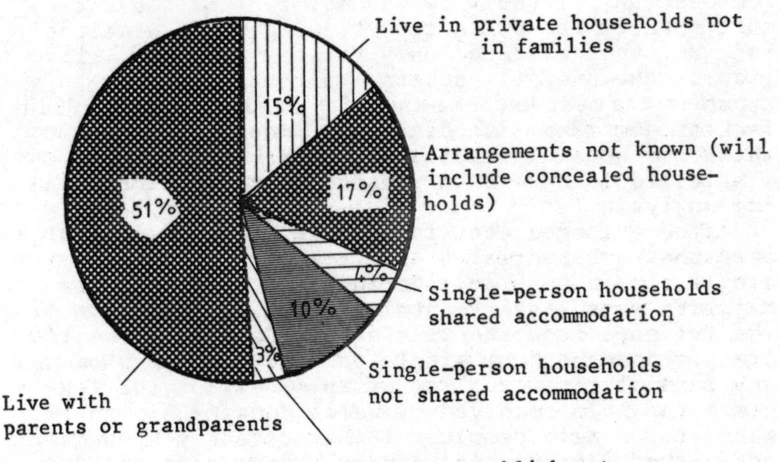

than 1 in 6 of the total number of single people of working age. Unfortunately, there is no reliable evidence on the proportion of single people who would like a home of their own. For example, the Sharers Survey (2) commissioned by the Department of the Environment only attempted to assess the preference of concealed married couple and lone parent households for independent housing. There are a number of studies which confirm the aspirations of homeless people in hostels and other forms of emergency accommodation for their own house, flat or bed-sit - most notably the DOE's report 'Single and Homeless'. (3) But this evidence only relates to a small fraction of the overall numbers - although clearly to a group in the most extreme severity of housing needs. The lack of comprehensive data underlines the degree to which the housing needs and requirements of most single people have been ignored in conventional housing analysis.

The evidence that is available, however, indicates that substantial - and growing - numbers of single people want to live independently, and the majority want a self-contained home of their own. The evidence from the Housing Aid Centres shows increasing demand from single people, most of whom lack any form of tenancy. For example, 41% of the requests for help received by Welsh Housing Aid in 1984 were from single people. It has recently been estimated that 276,000 single people were registered on housing waiting lists in England and Wales in 1984, (4) representing 27% of the total population of council waiting lists.

It is clear, however, that many single people who would like council housing are discouraged from applying, and in too many instances actually are barred from applying. The CHAR survey of local authority housing policies found a tangle of restrictions imposed on registration and eligibility for council housing. Of the 236 authorities which provided copies of the 'Section 44' leaflet (the information on allocation policies required by the 1980 Housing Act), 188 included some restriction on eligibility for even registering on the waiting list. Many authorities operated restrictions which prevented some single people being allocated housing, including age limits and residence qualifications. At the extreme are some authorities who will not even consider single people for council housing until they reach retirement age - including Bournemouth, Hove and Worthing. East Yorkshire District Council allow single people to register on the waiting list at 30, but will not offer accommodation until they reach retirement age - a minimum wait of 30 years! (5)

The evidence on local authority housing stock from the survey showed that most councils had fewer than 10% of total dwellings that were appropriately sized for single people (excluding accommodation specially designated for the elderly), although single people represented more than 20% of the total expressed demand - and in more than a quarter of authorities over 30%. (6) Very few councils had developed any accommodation to meet distinctive demand from single people, including purpose built single person dwellings, shared housing or furnished flats. The irrefutable conclusion is that the majority of local councils have effectively ignored the housing needs of single people. This attitude is reflected in barriers to application, discriminatory allocation policies and inadequate supplies of appropriate housing. The cumulative effect is that only 7% of all local authority households currently comprise single adults of working age.

The Conservative Government has single mindedly promoted owner occupation as the preferred housing tenure. The growth of home ownership has been feulled by massive tax relief and other benefits, including generous discounts to council tenants buying their homes. Yet few single people have been able to take advantage. Mortgage payments in the early years are especially onerous for those dependent on a single income, and impossible for those on low wages or unemployed. The 'right to buy' discounts are not available for those excluded from council accommodation. And in any event there are many single people for whom owner occupation is not necessarily appropriate, even if it were practicable.

The traditional alternative for many single people, of course, has been the privately rented sector. Single person households are still disproportionately represented in this sector, especially in furnished tenancies where 40% of all households are single people living alone. However, the long decline of private landlordism means that a number of new lettings from this source continues to become smaller. In 1914 90% of all households were rented from private landlords. By 1945 this had fallen to 60%, and today is less than 13% (and only 10% if housing association tenancies are excluded).

In itself, this decline is no cause for mourning. The private rented sector has always contained the greatest concentration of unfit, illequipped and badly maintained properties. Private tenants are most likely to live in overcrowded and multioccupied accommodation, to experience harassment and illegal eviction and to lack security of tenure. The erosion of Rent Act Protection, both through legislative changes and landlord exploitation of loopholes, means that a very high proportion of lettings are now unprotected.

The problems arise because no satisfactory alternative has been provided for those households who would previously have looked to the private rented sector. Tenants rented from private landlords through necessity, not choice. However, the disappearance of even this option means that large numbers of single people are unable to find any form of independent accommodation. As a result they are forced to turn to insecure, temporary accommodation.

Hostels, Night Shelters and Resettlement Units

According to the 1981 census there were 51,099 people living in hostels and common lodging houses in Great Britain. Whilst some of these are modern hostels with satisfactory standards, a large proportion are old institutional hostels and night shelters with the most intolerably primitive conditions. These include large communal dormitories or cramped cubicles, inadequate fire precautions, spartan toilet and washing facilities and authoritarian management regimes.

The traditional assumption was that single homeless people preferred to live in these types of hostel. It was argued that they would not be able to cope with a home of their own, and that they liked the anonymity and lack of responsibility of hostel life. The most authoritative repudiation of this belief is the DOE research 'Single and Homeless', which found that 85% of people who were homeless would prefer to live in a house, flat or bed-sit. One third would probably require some form of supportive housing, the majority no more than sensitive housing management. Moreover, the conclusion of the researchers was that these preferences were realistic. People live in hostels and bed and breakfast accommodation not because they want to, but because there is nothing else.

During recent years a number of successful compaigns have achieved a commitment to close some of the worst institutional hostels and replace them by a range of good standard housing. These include the action in Liverpool for the closure of the notorious Unique hostel (following a control order served by the city council); the programme for the replacement of the Camberwell Resettlement Unit in London; the takeover by three local authorities of the massive Rowton hostels in Camden Town, Vauxhall and Whitechapel; the replacement of the Manchester Night Shelter (and plans to replace also the two large hostels for homeless women and men run by Manchester Housing Department); and the control order served on the Princes Lodge hostel in Limehouse.

Since 1980 the Government has supported some of these initiatives through substantially increased funding for hostels developed by housing associations and voluntary agencies. More than 11,000 bed spaces have been approved by the Housing Corporation for this type of supportive housing project - mainly in small, good standard developments.

In January 1985 the Government also announced plans for the closure of a further 8 of the most dilapidated resettlement units, and have invited voluntary organisations and other bodies to put forward proposals for providing alternative arrangements. The aim is to close those units by March 1988. They also propose drawing up a timetable for the closure of the other 7 units outside London in the light of the proposals that may be made for their replacement.

The experience of hostel replacement programmes has demonstrated that residents can move successfully into more independent accommodation. It is clear that there needs to be a range of emergency, supportive and independent accommodation - and it is vital to provide preparation and active assistance to residents through the process of rehousing (through the work of 'resettlement' or 'home maker' teams).

There are, however, three key limitations to the work done so far towards the replacement of hostels. Firstly, funding has been channelled almost wholly through the voluntary sector and only minimal contributions have been made by local authorities. There are a few exceptions - notably the developments in Liverpool and the plans of four London boroughs to provide new direct access accommodation as part of the Camberwell replacement scheme - but in most areas of the country local authorities have left all this work to voluntary agencies. One serious consequence of this is that the organisations providing the new hostel accommodation have had little access to long term permanent housing, and residents have been trapped for far too long in projects intended only to offer a short term stepping stone to independent accommodation.

Secondly, the majority of the old institutional hostels have been primarily used by white men. There is now mounting evidence that black people and women also experience severe problems of homelessness, but typically rely on different forms of temporary shelter. Women's homelessness tends to be 'concealed', staying temporarily on the floor or sofa of friends or relatives - or remaining in intolerably oppressive domestic situations. Black people are also more likely to resort at least initially to emergency accommodation with friends, to use short life housing or to live in bed and breakfast accommodation. The reasons for this are complex and the experience of homelessness amongst women and black people has been neglected both in research and most conventional accounts. The causes include the much smaller provision of hostel accommodation for women and the racist attitudes experienced in some large old hostels. The consequences, however, have been clear: the concentration on the replacement of old institutional hostels has neglected the distinctive needs of women and black people, and thus represents yet another example of institutional racism.

Thirdly, despite the progressive initiatives in some areas, the rate of closure of beds in the old hostels has outpaced the development of new provision. Most particularly in London there has been a substantial fall in bed spaces in the traditional direct access hostels, and data now regularly collected by the GLC from all these hostels shows that almost every one is full to capacity every night. And this is happening despite the appallingly low standards in some of these institutions and the authoritarian management regimes. These grim statistics simply underline the evidence on the desperate shortage of more satisfactory alternatives.

Commerical Bed and Breakfast Accommodation

One consequence of the lack of access to satisfactory rented housing and the run down of old hostels has been the growing use of temporary bed and breakfast hotels by homeless single people. Statistics published by the Department of Health and Social Security show that the number of supplementary benefit claimants in board and lodging accommodation rose from 49,000 in 1979 to an estimated 139,000 in 1984 (Table 1). This total includes claimants in hostels and common lodging houses, but the numbers of people in this type of accommodation rose only from 24,000 to 35,000 between 1979 and 1983 (Table 2). It can be broadly calculated, therefore, that the numbers of claimants in bed and breakfast over the 5 years from 1979 to 1984 rose from 25,000 to 100,000 - a fourfold increase.

Table 1: Claimants in Ordinary Board and Lodging

Year	Numbers	Increase	Average Payment £pw	Increase	Expenditure £m	Increase
1979	49,000		20.40		52	
1980	55,000	12	26.30	29	76	46
1981	69,000	25	32.15	22	115	51
1982	85,000	23	37.80	15	166	44
1983	108,000	27	48.05	27	270	63
1984	139,000	29	52.35	9	380	41

Table 2: Estimated Numbers of SB Claimants Living in Hostels Etc. and as Boarders

Year	Hostels and Common Lodging Houses	Boarders (Thousands - GB)
1979	24	25
1980	24	31
1981	31	38
1982	31	54
1983	35	75

Source: Annual Statistical Inquiry (Hansard Col. 662 December 14, 1984).

According to the DHSS Statistical Inquiry more than 95% of claimants in bed and breakfast are single people.

This increase in the use of bed and breakfast has taken place not only in London, but in cities and towns throughout Britain. And contrary to some popular beliefs and media stories, conditions in most B and B establishments do not offer a life of luxury on the dole but condemnation to endure insecure, squalid and overcrowded accommodation. Evidence gathered by the Houses in Multiple Occupation Group in its 'Campaign to End Bed-Sit Squalor' has documented appalling examples of exploitation by landlords. (7) In a recent report the Institution of Environmental Health Officers confirm their belief that more than 80% of all multioccupied properties in England and Wales fail to meet satisfactory standards, and state that 'few local authorities have a comprehensive approach or a coherent policy'. (8)

In those areas where authorities have started systematically to inspect such properties and enforce adequate minimum standards, their findings have shown how poor conditions typically are. In Bristol, for instance, there are an estimated 340 commercial bed and breakfast establishments, used by up to 8,000 homeless single people. Following a policy decision by the City Council in 1984 to crack down on substandard premises, officers from the Environmental Health Department began a programme of inspection. Of the first 74 properties visited, 99% required remedial action.

The London Borough of Camden completed a pilot survey of 46 board and lodging premises used by homeless people in the Kings Cross area. Every one of 46 establishments was found to be overcrowded, all were lacking adequate catering facilities and 45 had inadequate amenities, especially toilets and bathrooms. Subsequently a control order has been served on one of these properties - the Spencer Hotel - because conditions were so appalling that they constituted a risk to the health and safety of the homeless people living there. In November 1984 Mrs. Karim, a homeless Bengali mother, and her two young children died in a fire in a bed and breakfast 'hotel' in Gloucester Place, Westminster, after being placed there by Camden Council Homeless Persons Unit. The evidence clearly shows that neither Westminster City Council, the GLC fire authority nor Camden Council had taken action to ensure effective fire safety for the people living there.

Yet the Government's response to these scandalous abuses is not to insist on proper standards or new forms of protection for tenants or improve access to decent rented housing. Instead the Government has simply introduced draconian new regulations to limit benefits to supplementary benefit claimants in board and lodging. In November 1984 proposals were published to introduce lower ceilings on the maximum charges that could be met by the DHSS and to restrict the eligibility for younger people to claim benefit. The Social Security Advisory Committee received an unprecedented barrage of submissions from individuals, voluntary and statutory organisations on the proposals, and their report to the DHSS was a scathing indictment of the Government's proposals.

Yet the regulations now introduced from 29 April 1985 are even harsher than those originally put forward. In addition to the new ceilings set for each local area (mostly considerably lower than the previous levels), there are now time limits on the periods of time for which young people aged under 26 can claim benefit in the same area. With the exception of those young people in an exempt category (e.g. those under 19 without a parent), benefit for board and lodging will only be paid for a maximum of 8 weeks in London, and a few other major cities, and as little as 2 weeks in some areas (mainly seaside resorts).

The justification given for these restrictions is that some young people have allegedly been taking 'holidays on the dole' in places such as Margate, Bournemouth and Newquay; that if younger claimants have not found employment after a given period whilst staying in board and lodging they should move on to another area; and that young people are deliberately staying in expensive bed and breakfast when they could return to their parental home. The evidence submitted to the Social Security Advisory Committee (SSAC) by more than 500 organisations refuted these arguments decisively. The Government themselves were forced to admit that evidence of abuses was largely anecdotal. The grim experience of young people looking for work is that jobs are simply not available. And the overwhelming majority of those staying in board and lodging accommodation do not have a parental home to which they could reasonably be expected to return. They are staying there not through choice, but because they are homeless.

The case against the new regulations was put succinctly by the SSAC in their report to the Secretary of State:

> We think the major problem with the revised proposals is the risk of creating a class of rootless young people, unable to find permanent accommodation in one place, unable to find a job and obliged by benefit rules to move around

> constantly. We do not believe it can be assumed that adequate alternative accommodation is open to claimants under 25 either in the public or private sector, or that permanent residence with parents or friends is an option which is realistically available. Benefit policy has to take account of the fact that even though board and lodging accommodation may be expensive and unsatisfactory, it is frequently the only real chance of housing the claimant has. We think the root problem here is a housing one, not a supplementary benefit one. (9)

The primary motive for the Government's action is to save money. The cost of supplementary benefit payments to claimants in board and lodging has risen very sharply over the past five years, from £52 million in 1979 to an estimated £380 million in 1984. The reasons for this, however, were primarily the worsening shortage of accessible rented accommodation. It cannot be denied that paying the cost of bed and breakfast to commercial landlords is expensive and wasteful. The evidence shows that it would not only be much better, but also cheaper, to enable local councils and housing associations to provide permanent rented housing. The central accusation against the Government is that the new regulations are simply a response to the symptoms. They wholly fail to tackle the causes of increasing homelessness.

Blaming the Victims

This analysis of homelessness amongst single people shows that the causes are structural. Contrary to many traditional assumptions, the explanations for homelessness are not to be found in pathological failures of individuals but in a housing system which limits access to satisfactory secure housing, especially for people on low incomes. The difficulties are compounded for those who experience additional handicaps, such as recent discharge from prison.

Pressure to develop progressive initiatives is constantly hindered by the use of labels which stigmatise homeless people. Terms such as 'dosser', 'vagrant', 'young drifter' carry the implication that people have brought their homelessness on themselves. This stereotyping is aggravated by the media images derived from the most visible homeless people

most frequently the 'cardboard city' of
under the arches at the Embankment in Lo
most homelessness is invisible and unmea
extent is unknown because there is no st
ponsibility to ensure that all homeless
somewhere to live. It is hidden becaus
less people are not on the streets - or
institutional hostels - but sleeping on the
or sofas of friends, living in insecure bed and
breakfast or squatting in empty properties.

The Alternative Approach - A Housing Policy for Single People

The starting point for an alternative approach is the recognition that people are homeless not because they have problems, but because there is a housing problem. The first priority must be to expand the provision of social rented housing. The number of new homes being built by the public sector has fallen from the level of 150,000 through the mid-1970s to fewer than 50,000 homes a year. Unless this downward trend is reversed, it will be impossible to eradicate the desperate shortage of satisfactory rented housing.

But resources alone are not enough. The record shows that most local authorities have totally failed to recognise the demand for independent housing for single people of working age. The CHAR 'Blueprint - Local Authority Housing Policies for Single People' (10) sets out a comprehensive framework of policies for action by local councils. This must include a proper assessment of the demand, the removal of all arbitrary barriers to access and the provision of a range of housing. Whilst the majority of single people simply want an ordinary self-contained flat, there is also the need for furnished accommodation, different forms of supportive housing and the provision of good standard emergency or 'direct access' accommodation run by the local council. All this implies a more flexible and sensitive approach to housing management and allocation. For example, young people moving into their first tenancy or people who have lived for years in institutions may require assistance in preparing to take on the flat, in obtaining furniture, in the connection of services and planning for the payment of bills.

It is also essential that the 1977 Housing (Homeless Persons) Act is extended to cover all home-

.. single people and childless couples. When the .ill was introduced into Parliament it was always envisaged that the categories of people designated to be in 'priority need' would be extended over time. It is now approaching ten years since the legislation was enacted, and still no progress has been made. The need for this became more urgent with the new DHSS board and lodging regulations. However inadequate, the possibility of obtaining bed and breakfast accommodation has been available as the last resort for unemployed people without a home. The lower cost limits and restrictions on young people mean that even this option has now disappeared for thousands of homeless claimants. Local authorities should be under a statutory duty to secure accommodation for anyone who is homeless.

In the short term action is urgently needed to enforce minimum standards of fire safety, overcrowding, repairs, amenities and management in all multi-occupied hostels, bed-sits and bed and breakfast lodging houses. The Housing (Houses in Multiple Occupation) Bill, promoted by the 'Campaign to End Bed-Sit Squalor' has already gained widespread support from an exceptionally wide range of tenants organisations, professional associations, local authorities, church groups and many more. The Department of the Environment themselves are carrying out research into the conditions in HMOs, the adequacy of existing legislation and the potential need for mandatory duties. The evidence of those pressing for new laws is that local authorities must have a clear legal duty regularly to inspect all multioccupied properties; faster and more effective procedures for ensuring that action is taken to enforce minimum standards; and obligation to ensure that any residents displaced by remedial action (or the victims of landlord harassment) are given satisfactory alternative accommodation; and that residents must have proper remedies in default of action.

Security of tenure is also essential - both for its own sake in giving tenants protection against the threat of arbitrary eviction, but also in enabling the enforcement of rights to obtain decent standards. The loopholes in the Rent acts allow landlords with increasing ease to let accommodation at high rents and without security, and there is the distinct possibility that the Government will introduce legislation to exempt all new lettings from Rent Act protection. There is no evidence that such measures increase the supply of accommodation - but a powerful body of experience which shows that insecurity results in intolerable experiences of landlord neglect, intimidation and unjustified evictions.

The basic principles should be that there is the right to a safe, secure and satisfactory home for every member of the community. And that must include the freedom for single people to live independently - young people leaving home, people coming out of hospital or other institutions, people leaving relationships - or those who simply choose to live on their own. Some will want support of one form or another, many will not. Yet all should have entitlement to adequate housing.

This programme of action clearly involves a massive transformation of current legislation, policy and the allocation of resources. It can only be achieved through Government support. Yet there are changes that can be made now, by local councils and by other agencies. There are opportunities to develop plans for the replacement of the old resettlement units and other institutional hostels and night shelters. Local authorities do have powers to act on unsafe and defective multioccupied hostels, bedsits and bed and breakfast hotels. Local councils and voluntary agencies can develop more comprehensive ranges of emergency, supportive and independent housing to meet the needs of single people. Plans can be worked out now - through consultation and collaboration with those who need the better provision - and be the focus for pressure and campaigning.

Finally, attitudes can be changed. It is not simply the lack of resources and the inadequate legislation, but blinkered and prejudiced ideas that have stigmatised homeless single people and condemned them to intolerable conditions. By developing a different analysis of the causes and an approach which recognises the legitimate aspirations of single people to satisfactory housing, it may be possible to advance the momentum for change.

NOTES AND REFERENCES

(1) S. Venn, Singled Out, CHAR, 1985. Estimates derived from OPCS Census 1981, Housing and Households, England and Wales.
(2) OPCS, Shares Survey (unpublished).
(3) M. Drake, M. O'Brien and T. Biebuyck, Single and Homeless, HMSO, 1982, p. 86.
(4) S. Venn, op. cit., p. 25.
(5) S. Venn, op. cit., p. 25.
(6) S. Venn, op. cit., p. 36.
(7) Houses in Multiple Occupation Group, Camp-

aign to End Bed-Sit Squalor, London, 1984.
(8) Institution of Environmental Health Officers, Houses in Multiple Occupation, London, 1985.
(9) Social Security Advisory Committee, Report on Proposals for the Supplementary Benefit (Requirements and Resources).
(10) Blueprint: Local Authority Housing Policies for Single People, CHAR, 1982.

Chapter Nine

HOUSING POLICY AND YOUNG PEOPLE

Peter Malpass

Housing policy in Britain is currently dominated by one objective above all others: the further expansion of home ownership. The owner occupied sector which has been increasing steadily for many years, has forged ahead since the Conservatives gained power in 1979. Survey evidence all suggests that this growth is consistent with a strong and growing consumer preference, and that the desire for home ownership is stronger among younger rather than older age groups. (1) On the face of it, then, the Thatcher Government is pursuing a housing policy that is both effective and popular.

However, closer examination soon reveals that the situation is neither so simple nor so rosy, especially for the large numbers of young people forming new households and seeking to establish independent homes of their own in the next few years. The Thatcher Government's reliance on market provision causes serious difficulties for young people setting up new households and seeking to establish independent homes. Young people also tend to be more mobile than older households and are therefore likely to encounter problems of access to decent housing repeatedly in a relatively short period as they enter and reenter the market. In the search for suitable accommodation young people are at a disadvantage when compared with older groups because not only do they move often but they tend to have more limited resources in terms of both savings and disposable regular income which restricts their freedom of choice in the housing market.

Access to council housing can be just as problematic, though for different reasons. Here the question is one of housing need as measured by the various sets of criteria used by individual authorities, and it is important to remember that councils are under no obligation to give priority to young people as such. In the past council housing was seen very much as family housing, with the emphasis in allocation schemes heavily on couples with children. Childless couples were expected to find accommodation in the private sector, and the continuation of this tradition can be seen in the way that single people and childless couples are excluded from the priority groups for rehousing under the Housing (Homeless Persons) Act of 1977.

The question of access is of particular importance for young people and can be seen as the basis for regarding young people as a group with a distinct housing problem. This is particularly relevant at the present time because of the way in which current policies are making access to housing more difficult for a growing proportion of young people. However access is only part of the problem and once housed young people especially if they have small children, may encounter difficulties in meeting the running costs of their accommodation. This is a problem shared with low income people of all ages, and it is important at the outset of a discussion which focuses on the housing problems of young people to establish that there are clear links with other groups. It is a mistake to see the housing problems of young people in isolation.

What links the various groups most at risk is low income and in many cases the associated difficulty of raising loans for house purchase or repair.

Like unemployment the housing problem mainly affects the working class, and in particular groups within the working class, such as those with low levels of skill in relation to the contemporary labour market, and those who are excluded from the labour force, including the elderly, disabled, many school leavers and women with young children. Once it is appreciated that class and labour market positions are the key connections with other age groups this also draws attention to the fact that young people are not a homogeneous group. Just as class links people across age groups in their common experience of housing and other problems so it also divides people within age groups. It is analytically essential to recognise that young people are not all in the same degree of difficulty in obtaining decent housing. For young people with secure, well paid jobs and inherited wealth, (2) there is effectively no housing crisis.

However fewer young people are likely to be in this personally happy position in the next few years.

A combination of demographic, economic and policy trends is set to exacerbate the housing problems of young people. It is convenient to present these as two pairs of conflicting factors. First, the number of young people in the age range 20 to 24 is increasing and will remain high until the end of the decade, reflecting the peaking of the birth rate in the mid-1960s. It is during their early twenties that most people embark upon the formation of separate households, although in recent years more people below the age of twenty have been setting up on their own, and if this continues it will only add to the demand for housing. The official expectation is that by 1991 there will be over 700,000 more households in England and Wales than in 1986. (3) Associated with the growth in the number of young households is an anticipated increase in the number of births until the early 1990s. (4) More new households, with more of them containing young children, seems certain to lead to increasing desire for separate accommodation and reduced willingness to continue sharing with relatives. But this need may not be met because of the low rates of building over the past few years and the problems involved in expanding output in the short term. There is, then, a conflict between growing need for housing amongst young people and the low level of investment at the present time.

Second, the economic recession has hit young people particularly hard, in the form of low wages, job insecurity and widespread unemployment. There is no immediate prospect of a sustained decline in the level of unemployment amongst young people, and it may even increase further. This means that many young people find themselves in a very weak position in the housing market and totally unable to share in the general aspiration for home ownership. The economic weakness of so many young people as consumers in the housing market is of course directly at odds with government policy which emphasises market provision, especially home ownership, and minimises public sector provision.

Restructuring the Housing Market

To understand the predicament in which young people find themselves in the housing market today a historical perspective is required. The point of this is to show that long term changes in the housing market, quite unrelated to the particular policies of the Conservative Governments since 1979 have made it more difficult for young people, and others on low incomes, to obtain access to private housing at a price they can afford.

Widespread home ownership is an invention of the twentieth century and only since the late 1960s have more than half of all households in Britain been owner occupiers (in Scotland owner occupation remains below 40% even now). In 1914 about 90% of households were private tenants and only 10% were owner occupiers; council housing was barely measurable in percentage terms until the 1920s. Since the First World War private renting has been in almost continuous decline and now constitutes only about 10% of the total stock, a figure from which it is unlikely to fall much further, while owner occupation has expanded to embrace almost two thirds of all households. As far as the private housing market is concerned, then, the last sixty five years represent a period of transition from renting to owning. This transformation of the market has been encouraged by successive governments, both Conservative and Labour, since the early 1950s, but it would be a mistake to attribute the change entirely to the influence of housing policy.

Home ownership expanded first and fastest amongst the better off, including better paid skilled workers. The less well off remained in the still dominant private rented sector because it was much cheaper to rent than to buy. For many years after the start of the restructuring of the housing market private rented housing continued to provide a large pool of cheap accommodation for the less well off, although much of it was also old and of poor quality. Council housing was, by contrast, newer, of better quality and more expensive on the whole, though still cheaper than owner occupation. It is only relatively recently, within the last twenty five years, that owner occupation has drawn in more and more working class families, and the growth of low income home ownership represents an important recent development in the housing system.

The significance of this from the point of view of low income households, and also governments such as the present one which seeks to build its housing policy around the further expansion of home ownership, is that the transition from a predominantly rented market to a predominantly owned market represents a shift towards higher entry costs. The costs involved in becoming a home owner are inherently higher than those of becoming a tenant, and therefore, unless policies can be devised to deal with the problems, home ownership is an expensive, inappropriate and even unattainable form of tenure for low income families. Whatever the long term advantages claimed for home ownership, if young people cannot afford the entry costs then it offers them nothing but frustrated ambitions.

Home Ownership and the Entry Costs Barrier

Ideally young people require housing with low entry costs, partly because of the difficulty of saving a large deposit and partly because mobility increases the burden by multiplying transaction costs (i.e. legal fees). Whereas rented housing is cheap to enter and leave because there are no legal fees and no deposits (or deposits which are relatively small), owner occupation is much more expensive. Despite the growing availability of high percentage mortgages most first time buyers find it necessary to save a substantial deposit; in 1984 the average deposit was about 18% of the purchase price, or in cash terms an average of £4,840. (5) It is important to stress that these are averages and obviously there are many dwellings on the market at below average price, closer to the purchasing power of lower income households. However, as will be discussed below, down market purchasing has its own potential drawbacks.

Saving a deposit equivalent to perhaps a year's net income in order to purchase a house is a daunting task, and is much more difficult where the income leaves little scope for regular saving. However, saving the deposit is only the first hurdle, and even if it can be avoided by obtaining a 100% mortgage the second hurdle is only made higher, for it is the problem of repaying the loan. The higher the percentage of the purchase price covered by the mortgage the higher is the monthly repayment. Borrowing money is much easier than repaying it and young first time buyers who take on the maximum mortgage offered by the building society can find themselves paying out a very high proportion of their take home pay, especially if the interest rate rises sharply soon after they make their purchase. People on higher incomes are better placed to cope with these problems because it is easier to pay a higher proportion of a high income on housing and still have enough left for other needs.

In inflationary conditions such as have prevailed in Britain throughout the period since 1945 owner occupiers face the maximum burden in repaying their mortgages in the early years of the loan. This is because, apart from changes in the interest rate from time to time, the repayments are pegged to the money value of the mortgage. Over time, therefore, assuming that wages rise in line with prices, the mortgage repayment constitutes a diminishing proportion of take home pay. The prospect of long run advantage is, unfortunately, small comfort to young people who often find that the period of maximum housing costs coincides with the reduced earning power and extra costs of family building.

Owner occupation is, then, relatively very expensive to enter and mortgage repayments can be highly burdensome in the first few years. Just when the burden begins to be reduced depends on the rate of inflation which raises the interesting point that while would-be owners have an interest in a low rate of inflation, recent purchasers have a strong interest in the continuation of inflation, the faster the better in terms of the erosion of mortgate repayments. Now the present Government has made much of both its determination to reduce inflation and the promotion of owner occupation. To the extent that it succeeds in the first, it reduces the appeal of the second.

Finally in this section, home ownership entry costs are rising over time in real terms. (6) This again shows the different positions of owners and aspirant owners, because owner occupation has been promoted on the basis that housing is an appreciating asset which throughout much of the postwar period has increased in value faster than other goods; to the extent that this occurs it increases the cost to first time buyers, who are mostly young people. It has been calculated that in 1975 the first year costs in real terms, for a first time buyer, were nearly four times the level in 1938. (7) The irony of this is, of course, that in 1938 there was a relatively plentiful supply of private rented housing (about 60% of the total stock) to which the less well off could turn. In the 1980s when more low income households are being drawn into owner occupation the costs are much higher.

All this points to the conclusion that the transformation of the housing market has made access to housing for young people at a price within their means more difficult, and the trend is increasing.

The Contemporary Housing Market

It is common for young people to start their housing careers in private rented accommodation, often furnished. Households under the age of 25 are much more likely to be private tenants than older people; in 1980 38% of household heads under 25 were living in private rented housing, compared with only 9% in the 30-44 age group. (8) Although the majority of private rented housing is let unfurnished this is mostly occupied by long standing elderly tenants who represent the vestiges of the old style private sector. Young people are much more likely to be found in furnished accommodation which, though suitable in the short term, is expensive, unsuitable and not what the great majority of young people desire for themselves in the long term. The furnished sector is characterised by very high mobility compared with other tenures, and can be seen as a tenure of transition through which young people pass on their way to more permanent homes. The role of the private rented sector thus remains important as a launching pad for housing careers in other tenures but it no longer offers young people housing for a lifetime.

In the long run the majority of young people aspire to owner occupation, but as the previous section has shown the system of financing imposes substantial entry costs. There are various responses to the problem. First, in order to maintain a demand for new housing builders, building societies and local authorities have devised ways of reducing the cost to first time buyers. It is not intended to describe these in detail but merely to refer to the fact that in principle there are two categories of approach. The cost can be reduced by financial mechanisms or by physical means, or some combination of the two. Financial mechanisms include low start mortgages, interest free periods, free conveyancing, provision of carpets and fittings - all designed to ease the burden of entry costs. Physical means simply refer to reductions in the size of houses offered for sale, the paring down of the quality of materials and finishes, and the omission of things like garages and central heating. There is clear evidence that since the mid-1970s and more especially since 1981, builders are producing a much higher proportion of dwellings with four rooms or fewer. (9) Builders can also reduce plot sizes and increase the density of development in order to keep prices down. What this means, of course, is that the rising real cost of home ownership is leading to young first time buyers having to tolerate lower standards.

And yet even when standards are reduced prices are still high in relation to the capacity to pay amongst a large proportion of young people. This can be illustrated by looking at two examples both of which were featured in a recent Nationwide Building Society newsletter. (10) First, a scheme in Oxfordshire, which was presented as helping first time buyers, involved a site developed jointly by the building society, a builder and the local council. The deal was for 63 one and two bedroomed houses to be sold on a 'cost-sale' basis which was claimed to be 10% below market price. However, the prices ranged from £16,049 to £21,305. Second, another scheme involving the building society, the same builder and Bristol City Council in which 126 dwellings, out of a total of 272 in the development, were offered for sale on the cost-sale basis. This time the prices ranged from £16,500 for a bedsitter to £39,000 for a 3 bedroomed house. Part of the deal involved the provision of 30 flats for renting, to be let by the city council to old people. The remainder of the dwellings were sold at full market price.

What these illustrations show is that even the schemes specially designed to reduce costs and help first time buyers result in the production of dwellings that are either too expensive or too small for the needs of young people particularly if they have children. In the first year a person buying a bedsitter in the Bristol scheme for £16,500 would pay £28 per week (after tax relief) in interest alone (assuming 100% mortgage and 12.75% interest rate). This is almost twice the average net rent for a council house.

Obviously there is a market for this kind of housing, but it does show the pressure that suppliers are under to maintain sales, and also the financial pressure that young first time buyers are under, even at income levels well above the minimum. Many young people are inevitably excluded from new housing, but they are sometimes able to buy in the second hand market. Here it is possible to find cheaper houses, usually pre-1914 dwellings, in inner urban neighbourhoods. However there is an important trade-off involved in going down market: the cheaper the house the more repair and modernisation it is likely to need. Young people on low incomes who can just afford the purchase price for a house at the lowest end of the market run the risk of being unable to afford unforeseen maintenance even if modernisation has been budgeted for. Such marginal and inexperienced purchasers are also perhaps more likely to economise on the entry costs by not paying for a full structural survey, and old houses that appear cosy and dry when presented for sale can easily conceal serious problems of dampness, dry rot, woodworm, etc.

The point here is that the comfortable, secure and financially rewarding suburban image of home ownership that is promoted before the public does not apply to marginal purchasers in the inner city. Young people seduced by the marketing, or forced into home ownership by the lack of any alternative, may find themselves financially overstretched and caught in a situation where to maintain the resale value of the house they have to carry out modernisation or repair that they cannot afford. Falling expenditure and rising value cannot be so readily guaranteed at this end of the market.

In terms of housing policy it is important to recognise that these problems exist for marginal purchasers, and that for the less well off home ownership is not an unequivocally satisfying or suitable tenure. At present young people who would in earlier times have gone into rented housing are being drawn into buying houses that would now be condemned if previous policies had continued. This represents a serious redistribution of the burden of old and obsolete housing from the state onto people who are least able to deal with it. In this sense, then, young people who are drawn into home ownership at the bottom end of the market are victims rather than beneficiaries of the policy of expanding owner occupation. It is essential in any assessment of housing policy and its impact on particular groups in the population to remember that home ownership is now a highly varied tenure, with very different patterns of cost and benefits at different levels of the market.

In order to respond to this situation and to sustain the market amongst low income households it is necessary to provide substantial financial assistance towards repair and modernisation. In April 1982 the Government announced an important development in the form of higher percentage grants for repair. For a limited period grants were raised to 90% of expenditure up to £5,000, and as a result there was an enormous increase in grant expenditure. The availability of the higher level grants was extended to March 1984, but since then grants, even at the lowest level, have become very hard to obtain in most areas as resources for local authority capital spending have been reduced by central government. (11)

A further blow to young people and others on low income was the introduction of VAT on building improvements with effect from June 1984. So after a promising move towards additional assistance for marginal owners the Government has put its policy into reverse.

Apart from the grants episode (which was not aimed at easing the route into owner occupation) the Government has introduced a series of low cost home ownership initiatives. (12) By far the most important of these, both in terms of the prominence given to it by the Government and the number of houses involved, is the sale of council houses at discount under the 1980 Housing Act. The introduction of the 'right to buy' gave council tenants of three years' standing entitlement to 33% discount from the market price and the right to a 100% mortgage on the sale price. After twenty years tenants were entitled to 50% discounts. In the three years 1981-83 over 429,000 dwellings were sold under the right to buy scheme in England and Wales. (13) The 1984 Housing and Building Control Act has now increased the maximum discount to 60% and reduced the qualifying period for the right to buy to two years. It also introduces the right to buy 50% of the house with the option to acquire the remainder in 12½% stages. This last measure is designed to appeal to low income tenants who cannot afford to buy the whole of the house at once. However, the importance of the right to buy is that it is most advantageous to tenants in middle age because they are much more likely to be entitled to the maximum discount. The sliding scale of discounts according to length of tenancy obviously biases the system against younger tenants and research on sales under the right to buy confirms that, 'the middle aged, skilled manual worker with a grown up family has been shown to be the typical council house purchaser.' (14) The right to buy is really of very little value to young people but it does represent an unrepeatable bargain for older people in the public sector and in this sense is very unfair in its treatment of different age groups.

There are six other initiatives intended to promote low cost home ownership and they can be dealt with quite briefly since, so far, they have made little impact and most of them are reworked ideas already in operation either locally or nationally. The first is what are called starter homes, conceived as a minimal dwelling that provides a first rung on the ladder of home ownership. The schemes referred to earlier in Oxfordshire and Bristol fall into this category, as do much more limited schemes modelled on third world projects in which the purchaser is provided with a site with basic services laid on and is left to build their own house. Second, improvement for sale, which means that local authorities can acquire old houses and carry out modernisation pending resale for owner occupation. Third, homesteading, where the local authority sells unimproved houses to be modernised within a specified time by the new owners. Fourth, shared ownership or equity sharing, in which the occupier owns part of the equity in the house and rents the rest from the local authority. Fifth, local authority guarantee powers, which enable local authorities to guarantee building society mortgages, the idea being to persuade societies to lend on properties that otherwise they would refuse. Finally, housing associations have been encouraged to embrace the expansion of home ownership in addition to their traditional role in rented housing.

It has been said that, 'Collectively these policies represent a varied and enterprising expression of concern to persuade both public agencies and individual households to recognise the advantages of home ownership. They represent a considerable investment of policy innovation and ingenuity.' (15) However, they have made a minimal impact on the housing market and have been virtually irrelevant to the needs of most young people seeking entry to home ownership. Just how irrelevant these measures are can be appreciated by remembering that whereas perhaps a few thousand dwellings have been provided for young people, <u>all</u> prospective and recent purchasers are affected by interest rate changes. A reduction in the mortgage interest rate of 2 or 3% would make much more difference to many more people than all these various initiatives. However, under the present Government mortgage interest rates have been driven higher than ever before, reaching 15% in 1979-80 and again in 1981. The impact of interest rate increases was such that the ratio of initial repayments to average earnings which stood at 22.1% in June 1978 rose to 35.8% by November 1979 as the interest rate rose from 9.75% to 15%. (16) This hike in the interest rate had a crippling impact on new and recent buyers and demonstrated the need for lower, more stable interest rates if home ownership is to continue to expand amongst the less well off and young people.

The Collapse of Council Housing

Since 1979 the demand for owner occupation amongst young people has been artificially amplified not just by the initiatives described above but by an attack of almost frenzied severity on council housing. There are three main components of this attack. First, as mentioned at the beginning, new building has been cut to the lowest levels since the 1920s as local authorities have faced a succession of substantial cuts in their permitted capital spending. This has had the effect of reducing the supply of dwellings for young people who cannot afford to buy. To make matters worse for the young the Government has encouraged local authorities to concentrate on building sheltered housing for the elderly (although this may release some family sized dwellings for reallocation to young people), and other 'special needs'. This is fully consistent with the Government's view that council housing should be confined to a residual role, catering for people who are unable to provide a profit for the private housing market.

Second, the introduction of the right to buy has led to a situation in which sales exceed new building, and after sixty years of growth the council sector is now in decline, both numerically and proportionately. This represents a historic turning point in the development of public housing. In the past sales were always outnumbered by new additions, usually heavily outnumbered, but no longer:

	Dwellings Completed By Local Authorities (England and Wales)	'Right to Buy' Sales In England and Wales
1981	49,411	97,055
1982	29,859	196,680
1983	30,024	135,895

The sale of council houses adversely affects people on the waiting list because it reduces the number of relets, especially in the longer term. It also means that, because houses with gardens are sold in much greater number than flats, and because sales are more frequent on the more popular estates, people on the waiting list have a more restricted choice of dwellings and areas. It means, too, that young people entering council housing now have less chance of securing the sort of dwelling that might encourage them to exercise their right to buy. It is important to remember that the right to buy is only a right worth having for tenants who can afford to exercise it and who live in dwellings that are worth buying. As the asset stripping of the public sector continues so young people entering council housing will find it harder to satisfy the second of these conditions.

Meanwhile as opportunities to enter council housing diminish, waiting lists are bound to grow unless new young households can be absorbed by owner occupation. The Government is clearly pursuing a deterrent policy; by reducing the supply of council housing it hopes to deflect demand into owner occupation. There are currently about 1,200,000 households on council waiting lists in England alone (17) and recorded homelessness has risen with every year since 1979. (18) Some hard pressed authorities have found that the homeless, for whom they have a statutory responcibility to provide, are absorbing a very high proportion of available accommodation, to the virtual exclusion of normal waiting list cases. In these circumstances young people with children have no chance of being housed unless they are prepared to accept the least popular dwellings and the least popular estates. Inevitably the most desperate end up in the worst estates and people with the weakest position in the market find themselves discriminated against in council housing too.

The third part of the attack on council housing has been the massive increase in rents since 1979. Between 1979 and 1982 rents were increased, on average, by 111%, mainly as a result of the withdrawal of subsidy. The 1980 Housing Act introduced a new subsidy mechanism for local authority housing which enabled the Secretary of State to withdraw subsidy in accordance with rent increases that he determined. In 1981 Michael Heseltine set a rent increase of £2.95 per week, and followed up with £2.50 in the next year. As a result most local authorities in England and Wales now receive no general housing subsidy at all. But what they do get is housing benefit. Essentially what has happened in the last five years is that there has been the completion of a process which began much earlier, of shifting away from general subsidies in the public sector to means tested assistance based on the incomes of individual tenants. (19) This is sometimes referred to as subsidising people rather than houses, but the importance of the change is that council tenants, on the whole, now receive assistance with housing costs as a form of means tested social security. Whereas the value of tax relief on mortgage interest increases with rising income, housing benefit tapers away as income rises. Coupled with big rent increases this is intended to dislodge the better off tenants, who have thoughtfully been provided with a convenient loophole, the right to buy. It is also designed to deter potential tenants and to encourage them into house purchase.

In considering the present onslaught on the public sector it is instructive to compare the different treatment of young and elderly people. The elderly continue to be the top priority group in the sense that not only is new building targeted on them but also the quality of new sheltered housing is very high. By contrast the needs of young people are not even recognised in public sector policy, unless they have children, and, as has been argued above, there is an increasing likelihood that young entrants will be allocated to the least desirable parts of the stock. In its almost obsessive promotion of owner occupation the Government has completely failed to make provision for the substantial numbers of young people who will require rented housing in the public sector. The underlying assumption seems to be that there are certain special needs for which the local authorities should provide and everyone else who falls into the category of general need should be provided for in the private sector. A policy of less eligibility applies to those general needs households who find their way into council housing: nothing should be done to make access easy for them and they should be given every incentive to leave.

Conclusion

The dominant theme running through this discussion is that there is a deepening housing crisis in Britain, affecting certain groups, including young people, more than others, and that the housing policies pursued by the Government are making matters worse rather than better. At just the time when the numbers of young people seeking accommodation are increasing, and when more of these new households require rented housing because of the recession and high unemployment, the Government has ruthlessly cut back the supply of council housing by reducing new building and promoting the right to buy. Although the right to buy has been of great benefit to established council tenants old enough to take advantage of the highest discounts, it has offered little to young people, especially those queuing up for council housing. At the same time, the Government's wider economic policies have made entry to home ownership prohibitively expensive for an increasing proportion of young people. The use of high interest rates and unemployment as tools of economic management combined to squeeze young people out of the housing market. In this sense the Government has pursued not just an anti-council housing policy but an anti-housing policiy.

Doctrinaire commitment to more and more home ownership has prevented consideration of changing housing needs and the inappropriateness of owner occupation for those on low incomes, particularly at the present time. In the 1950s and 1960s, when inflation, interest rates and unemployment were all low and economic growth seemed guaranteed, mortgaged home ownership was clearly attractive. The irony is that as this tenure has spread to a wider section of the less well off changing economic conditions have made it less certainly rewarding for such purchasers. At present changes in the savings market, including the breaking of the building societies' interest rate cartel, are tipping the balance of advantage towards the investor and away from the borrower. Proposed legislation to widen the scope of building society activity, making them more like banks, seems likely to exacerbate this trend, further reinforcing the view that the 1950s and 1960s were a golden age for home owners, and providing little comfort for young people embarking on their housing careers in the 1980s.

Housing policy is failing young people today because of a failure to recognise their needs, and also a polarisation between home ownership which is 'good' and council housing which is 'bad'. It is important to break out of this policy straitjacket and recognise that on the one hand home ownership has major drawbacks, and that the widespread preference for this tenure is not innate but a rational response to the prospect of financial advantage in the long run. At the same time, council housing, on the other hand, has been manipulated into a position where it is seen as less desirable than owner occupation, especially for people on higher incomes, but nevertheless it does have certain inherent advantages to offer.

It has been argued above that young people need

housing with low entry costs, which is exactly what council housing offers. The great strength of council housing now and in the future (which helps to explain the virulence of the current attack on it) is that the accumulated stock of houses built at much lower prices over the period since 1919 enables rents for new houses to be kept well below levels reflecting current building costs. Council housing is financed collectively and tenants in new expensive houses are effectively subsidised by tenants in older houses. This characteristic enables council housing to offer lower entry costs than the individually financed system of home ownership. In addition the size of the stock and its cost structure is such that low rents can be achieved without subsidy.

But, council housing does not offer the possibility of capital accumulation - tenants have nothing to show for their years of rent paying, nothing to leave to their children. A way round this might be to continue with a right for tenants to buy their houses but to introduce an obligation to sell back to the council when the owners wish to move. This would enable owners to acquire a capital sum and also preserve the stock of houses available for letting at rents within the reach of new young households.

Essentially what is being proposed here is a dissolution of the importance of housing tenure divisions. It should be recognised that council housing has a unique capacity to play an important role in housing young people and new building should be expanded considerably to provide for their needs. At a time in their life cycle when tenants feel that they can cope with the higher cost of home ownership it should be available in the form of a right to buy. Still later, in old age, such owners may prefer to become tenants again, to realise their capital or to avoid the worry of repair and maintenance responsibilities, and a right to sell back to the council should be established. This right to sell could also be extended to owner occupiers in the conventional housing market; it would be of great benefit to the elderly and to young people who for whatever reason found themselves unable to continue as home owners.

The point of this proposal is that home ownership and renting are most appropriate at different stages in the life cycle, with renting being best suited to young and elderly households. There are great benefits to be obtained by developing a housing policy which permits people to move back and forth between tenures as their needs indicate. Unfortunately, such flexibility seems remote at the present time.

NOTES AND REFERENCES

(1) Building Societies Association, Housing Tenure, 1983. See also C. Jones, The Demand for Home Ownership, in J. English (ed.) The Future of Council Housing, Croom Helm, London, 1981.

(2) A. Murie and R. Forrest, 'Wealth, Inheritance and Housing Policy', Policy and Politics, vol. 8, no. 1, 1980 pp. 1-19.

(3) Housing and Construction Statistics 1970-1980, HMSO, 1981, TAble 142.

(4) Social Trends No. 14, HMSO, 1984, p. 21.

(5) House Prices, Second Charter, 1984, Nationwide Building Society.

(6) BSA Bulletin no. 39, July 1984, Table 5, p. 18.

(7) Housing Policy Technical Volume Part 1, HMSO, London, 1977, p. 45.

(8) General Household Survey 1980, quoted in Building Societies Association, Housing Tenure, 1983, p. 7.

(9) BSA Bulletin no. 39, July, 1984. Table 6, p. 21. In 1974 1% of new houses had fewer than 4 rooms and 12% had 4 rooms; by 1983 the corresponding figures were 13% and 25%.

(10) Nationwide Housing, no. 5, Spring 1984.

(11) N. Fielding, 'The Next Step in the Improvement Grant Fox-trot', Roof, July-August 1984.

(12) S. Lansley, R. Forrest and A. Murie, A Foot on the Ladder? An Evaluation of Low Cost Home Ownership initiatives, SAUS Working Paper no. 41, University of Bristol, 1984.

(13) Housing and Construction Statistics, March 1984.

(14) R. Forrest and A. Murie, Monitoring the Right to Buy 1980-1982, SAUS Working Paper no. 40, University of Bristol, 1984, p. 39.

(15) S. Lansley et al, op. cit. p. 73.

(16) BSA Bulletin, no. 39, July 1984, Table 6, p. 18.

(17) AMA 1984, op. cit. p. 8.

(18) AMA Submission into the Inquiry into British Housing, 1984, p. 8.

(19) P. Malpass, 'Housing Benefits in Perspective', in C. Jones and J. Stevenson, The Year Book of Social Policy in Britain 1983, RKP, 1984.

Chapter Ten

CONCLUSION

Peter Malpass

In their different ways the various contributors to this book have examined the nature of the current housing crisis and its impact on both consumers and, to a lesser extent, providers of accommodation. Each chapter has identified a serious and worsening situation, the full severity of which is felt most acutely by some of the most vulnerable and least powerful groups in British society. In this final chapter the intention is, first, to pull together in summary form some of the themes emerging from the earlier analysis; second, to put forward some ideas about what kinds of policies are needed to alleviate current problems, and third, to reexamine the notion of crisis, recognising the important distinction between the private crises of homelessness, overcrowding, mortgage arrears etc., experienced by individuals and families, and the public crisis that can only occur as a result of political action to change housing policy.

Not So Much A Policy...

Since 1979 the Conservative Government of Margaret Thatcher has, in one sense, pursued a coherent and politically successful housing policy, based on cuts in public expenditure on the housing programme and the expansion of home ownership. However it has been suggested that this is merely a tenure policy, not a housing policy, (1) although it is possible to go further and to argue that the Government has been pursuing an anti-housing policy. The charge that the Government has merely a tenure policy is based on the centrality of the council tenants' right to buy, which is about transferring the ownership of existing dwellings, having little or nothing to do with the key housing policy issues of quantity and quality. Moreover, the sale of council houses is a policy aimed at providing benefits for people who are already well housed, doing nothing for those in greatest need. A key feature of current tenure policy is its unfairness, best illustrated by the imposition of a strictly means tested housing benefit scheme for tenants (a scheme which has been a target for spending cuts), while the inverted means test of tax relief on mortgage interest is defended despite escalating cost. In addition the tenure policy accusation rests on the observation that for the individual household the achievement of home ownership is assumed to represent the end of their housing problem. A theme running through a number of chapters in this book is that this assumption is deeply flawed.

The idea that the Thatcher Government has pursued an anti-housing policy goes beyond the attack on council housing to recognise the contradictions between housing and economic policy objectives. First, it is clear that the housing programme bore the brunt of the cuts in public expenditure in the early 1980s, (2) not for any carefully argued housing reasons but for political and economic reasons. Second, the use of high interest rates and high unemployment as tools of economic policy, at the same time as relying on further growth of home ownership, is highly contradictory. Economic policy has frustrated the achievement of housing policy objectives by raising the cost of mortgages. In addition, one of the main attractions of home ownership presented by Government spokesmen and women has been its financial advantages, yet high interest rates and low inflation seriously undermine or postpone attainment of these advantages. The people who have gained most out of home ownership are those who bought before the bouts of rapid house price inflation in the early and late 1970s, and those who bought before the periods of increase in the general rate of inflation. The defeat of inflation may help savers who aspire to home ownership, but recent purchasers have a clear interest in continued price and wage inflation to erode the real value of their mortgage repayments. When low inflation is combined with high interest rates then the Party of home ownership is clearly pursuing economic policies which conflict with its housing rhetoric.

The cuts in public expenditure on housing and further expansion of home ownership are two arms of one policy. However this policy has been shown in preceding chapters to be a major contributory factor in the creation of the current housing crisis. The key conclusions can be briefly stated:

(i) that cuts in public investment in housing are a short term expedient that stores up problems of greater shortage and disrepair in the future;
(ii) that the state of disrepair in both public and private housing is already very serious and rapidly getting worse;
(iii) that reliance on a policy of further growth in home ownership is inappropriate, especially in a period of high unemployment, because home ownership and low income are highly incompatible (in the absence of specific measures to reconcile them);
(iv) that it is clearly wrong to build a housing policy on the assumption that cuts in public investment will be made good by private sector investment.

Towards an Authentic Housing Policy

A point that was made with some force in the opening chapter was that housing problems could not be defined away as aspects of other problems such as poverty, old age, unemployment etc. This notion has also been implicit throughout the book. It follows, therefore, that what is required to solve the current housing crisis is a set of policies designed to confront housing problems as such. There is now a widespread political consensus in Britain that home ownership is the preferred tenure of the majority of the population, and that the long term role for council housing is to provide for those who must rent.

In devising a housing policy that is intended to meet the traditional objective of a decent home for every household at a price within their means (itself a very imprecise goal), (3) *any* government must face up to the fact that mass home ownership is not a cheap option in terms of its public expenditure implications, and that local authorities must be given the resources to carry out their agreed role satisfactorily. A major cause of current housing problems is the Government's refusal to recognise these requirements. What is happening at present is that, on the one hand, more and more people are being shoehorned into home ownership without the help of a financial framework and other support mechanisms geared to their needs, and on the other hand local authorities are denied the power to borrow the funds they need to build houses for people who require rented housing.

In the case of local authority housing three measures that would go some way towards improving the situation are i) release of accumulated capital receipts from council house sales, ii) a return to much more generous limits on borrowing for capital investment, and iii) reintroduction of some kind of subsidy on new building.

In 1985 local housing authorities have over £5 billion in accumulated capital receipts which they can use only slowly following the introduction of tighter rules in April 1984 and 1985. However even if this money were to be spent it would go nowhere near meeting the need for investment in new housing and refurbishment of the existing stock. It is therefore necessary to relax limits on borrowing to finance capital projects. The longer capital expenditure is held down the greater will be the amount that has to be spent in the longer term. The need for subsidy arises from the fact that the majority of authorities now receive no general housing subsidy under the 1980 Housing Act, and many, according to the Department of the Environment's way of looking at these things, are running notional surpluses on their housing revenue accounts. For sixty years new building by local authorities attracted automatic subsidy, but this is no longer the case. Authorities embarking on ambitious new building programmes at the present time (if it were possible) would have to eliminate their notional HRA surpluses before qualifying for subsidy. This means that the cost of new building would fall entirely on the housing revenue account, with important consequences for rents, which, of course, is a powerful disincentive to build.

For some local authorities the major issue is not so much a shortage of dwellings as a surfeit of defective and unusable system built dwellings constructed within the last twenty years. To help these authorities it might be necessary to allow them to write off debts on dwellings that have to be demolished, by transferring liability to central government.

Turning to problems in the owner occupied sector, a variety of responses are required to tackle different kinds of problems. One of the most urgently needed reforms is in the area of housing finance, where there are a number of issues to be resolved. Since 1979 the Conservative Government has carried out a thoroughgoing reorganisation of public sector housing finance but it has scrupulously avoided the issue of reforming arrangements in the owner occupied sector. As a result it has increased the amount of unfairness between tenures and failed to reduce the unfairness within home ownership.

In order to make home ownership work successfully for low income purchasers it is necessary to increase the level of assistance given towards both mortgage costs and repair, maintenance and modernisation costs. In particular it is important to extend housing benefit to owner occupiers (at present they qualify only for housing benefit in relation to their rates bills). The case for a unified housing benefit developed by the Supplementary Benefits Commission and its Chairman, David Donnison, in the late 1970s, presupposed the inclusion of owner occupiers, and a simultaneous reorganisation of the whole system of assistance for home owners. (4) Low income purchasers are the main group of housing consumers to suffer from the failure to take up the challenge of fundamental reform of housing finance. They are the people who obtain least help from the existing arrangements for tax relief on mortgage interest, which provides most help to those who need it least.

The reform of tax relief should be a high priority, yet Margaret Thatcher and Neil Kinnock seem equally unable to grasp its importance. Their commitment to its retention has led The Guardian to comment that, 'Mortgage tax relief and the mortgage interest rate itself have become absurdly exaggerated political totems.' (16th April 1985). The case against mortgage tax relief has been considerably strengthened by the restructuring of public sector assistance with housing costs. Now that tenants only receive help on a means tested basis it is much more difficult to justify the indiscriminate and inequitable distribution of the benefits of tax relief. Reform of mortgage tax relief could simultaneously produce an increased tax yield for the Treasury and much more assistance for owners in real need. In terms of both fiscal rectitude and simple equity amongst householders the case for reform is overwhelming. It is only short term electoral fetishism that prevents action. But it should be remembered that only a minority of home owners would be directly affected by a change (i.e. those with large, recently acquired mortgages), and transitional arrangements could be devised to minimise hardship.

On the question of assistance with repair, maintenance and modernisation costs the point is that low income purchasers who are already fully stretched financially by their mortgage costs are least well placed to keep their property in good condition, and most likely to live in old and dilapidated dwellings. Elderly people on low incomes also face great difficulties in this respect, even though they usually have no mortgage burden. Unless there is greater recognition of the problem and greater provision of assistance to tackle it, disrepair in the owner occupied sector seems certain to spread, possibly with important long term consequences for the popularity and survival of home ownership amongst people on low incomes.

There are several ways of tackling the problem. The availability of 90% repair grants in 1982-84 proved immensely popular and resulted in a huge increase in expenditure, although this was not always well monitored on the ground, nor was it efficiently channelled to those in greatest need. However this episode demonstrated the need for repair grants at a high percentage rate. Nevertheless there will be some owners, especially elderly people, who cannot face the prospect of arranging and enduring major repair and improvement work. An escape route for such people could be provided by introducing a statutory 'right to sell', the counterpart to the right to buy. There are already private sector schemes to provide elderly owners with access to their capital, and Anchor Housing Association operates the 'staying put' scheme for owners who sell their homes to the association and remain in residence but without the responsibility for upkeep of the property. The right to sell would extend the opportunity for elderly owners (and not necessarily just elderly owners) to trade ownership of their house (or part of the equity) in return for a maintenance and modernisation service provided by the local authority or a housing association.

Under the heading of the right to sell a variety of schemes could be introduced, to give owners a choice of equity sharing arrangements or outright sale of their property. From the owner's point of view the advantages would be access to capital and no further worries about the condition of the dwelling. From the local authority's point of view the advantages would be possibilities for converting large dwellings into flats, a more mixed stock of public sector accommodation and confidence that old houses were not deteriorating to the point of requiring very expensive works or even demolition. Alternatively, local authority direct labour organisations could offer a repair and maintenance service to the private sector for an annual premium or payment for each job. In the long run the right to sell could become a self-financing trading activity as local authorities returned houses to the open market after the death or departure of the occupier who exercised the right to sell.

The buying and selling of houses is an area in which local authorities could become much more actively engaged. The case for a municipal estate agency service is, first, that at present the costs of buying and selling represent a heavy burden for low income households and a barrier to entry to home ownership. Second, buyers and sellers often become involved in chains of transactions which can be disrupted if just one person pulls out. The local authority could offer a very much cheaper service, without risk of broken chains, by acting not as an agent between buyers and sellers, but by actually buying and selling houses itself. It would gradually acquire a 'bank' or stock of houses that had been bought on the open market and were offered for sale in the same way. Vendors could dispose of their dwellings quickly, simply and without the worry of the sale falling through at the last moment. On the other hand purchasers could choose from the stock offered for sale by the authority, knowing that each house had been professionally surveyed by council staff.

Low income buyers and sellers in particular could expect to benefit from a service of this kind. Struggling home owners unable to afford essential repairs would have a guaranteed market, and first time buyers would avoid the risk of acquiring a dwelling that soon revealed expensive faults which had been concealed or unnoticed before the sale was completed. Local authorities might expect to become much more involved at the bottom end of the housing market, buying rundown old properties and improving them for sale, or offering unmodernised dwellings for sale with a guaranteed improvement grant.

These are just some of the ways in which the current housing crisis could be alleviated; this list is not comprehensive or original, (5) and in particular it lacks reference to the plight of the homeless and problems in the private rented sector. (6) The problem of homelessness could be reduced if local authorities were given the resources to provide more accommodation, and required to accept as tenants a much wider range of households, including single people. For many years now there has been talk of local authorities providing a 'comprehensive' housing service, but in practice they have remained primarily oriented towards the provision of rented accommodation for families and the elderly. Given the spreading of housing problems in the owner occupied sector and the increasing need for some tighter regulation of parts of the private rented sector, perhaps the time is ripe for the development of a truly comprehensive service that would include the services suggested above.

Crisis, What Crisis?

Finally it is necessary to return to the notion of crisis itself, and to recognise that the proliferation of housing problems for individuals and families does not necessarily amount to a crisis from the point of view of the Government. To identify disturbing tendencies in housing conditions and to expose the deficiencies of the housing system, as this book has done, and to make suggestions about what should be done to improve the situation is ultimately irrelevant if the Government is not listening. It is important to set developments in housing in a wider context of social and economic change. There are, for instance, chilling similarities in recent housing problems and policies in Britain and the United States as right wing governments have responded in similar ways in attempting to manage the economics of the recession. (7)

On both sides of the Atlantic governments have used housing as a prime target for achieving the cuts in public expenditure that have become the new economic orthodoxy in the post Keynesian era. These cuts fall most heavily on groups who are on the margins of the economy (the unemployed, school leavers, the elderly etc) while the living standards of those remaining in employment are protected - in the area of housing it is convenient that tax relief on mortgage interest still does not count as public expenditure. The worsening housing problem is, therefore, like unemployment, a by-product of conscious policies, and to recognise and respond to these problems would involve changes in the central planks of economic policy.

A crisis is a turning point, a time when changes

have to be made in response to pressures and tensions building up to a peak. The private crises of individuals and households do not amount to a public crisis requiring a change in policy unless those personal experiences are translated into political pressure. Recent British experience suggests that it is very difficult for the working class and opposition groups in general to put enough pressure on the Thatcher Government to bring about changes in policy. It is clear that moral indignation about the plight of the homeless or families living in squalid conditions or old women freezing to death in unheated homes is no longer an effective lever (if indeed it ever was), given a Government that is prepared to 'tough it out'. In this sense, then, it seems that the housing crisis has not yet arrived. However, its approach is made closer by growing contradictions in policy which will become increasingly obvious and more difficult to avoid as the decade continues.

NOTES AND REFERENCES

(1) D. Donnison and D. Maclennan, 'What Should We Do About Housing?', New Society, 11th April 1985.

(2) First Report of the Environment Committee, Session 1979-80, HC714, HMSO, London, 1980, P.V.

(3) See D. Griffiths, in Labout Housing Group, Right to a Home, Spokesman, Nottingham, 1984, p. 21. Also P. Davidoff, in C. Hartman (ed.) America's Housing Crisis, Routledge and Kegan Paul, London, 1983, Ch. 6.

(4) D. Donnison, The Politics of Poverty, Martin Robertson, Oxford, 1982, p. 188.

(5) This section draws heavily on discussions with Alan Murie and Ray Forrest, but these ideas, and others, are also currently being considered in forums such as the Labour Housing Group, op. cit.

(6) C. Whitehead, M. Harloe and A. Bovaird, 'Prospects and Strategies for Housing in the Private Rented Sector', Journal of Social Policy, vol. 14 no. 2, 1985, p. 151-174.

(7) C. Hartman, Chapter 1 and C. Hartman (ed.) op. cit.

INDEX